Temperature Perturbations

A scientific analysis of the planet Mars and its influence on terrorism, rainfall, and stock market crashes

Anthony of Boston

This book is divided into 3 different academic papers which uses credible scientific analysis to explain how the planet Mars exerts influence on earthly matters via its gravitational pull. This gravitational effect influences temperature perturbations, which in effect influences weather and human behavior. This allows us parlay scientific facts onto prediction models where correlation is near 100%. Consequently, we can conclude from the data that the correlation does in fact indicate causation.

The first two papers provide scientific reasoning and evidence that the data showing correlation between Gaza rocket attacks and stock market crashes to the configuration of the planet Mars in relation to the earth is evidence that there is a link between the physics involved at the astrophysical level, the meteorological outcome and its effect on the biological processes of earthly organisms, which manifest certain behaviors.

The third paper makes astute observations which posit a link between alignment of the moon and Mars with the timing of extreme rainfall events in the Middle East

Studies from 2014 and 2024 are referenced in this book to elucidate the scientific basis involved in the research. Both studies link the motions of celestial objects to weather variations and climate. Other studies link weather variations to human behavior. All of these can be traced back to the orbit of the planet Mars

Back in 2019, using rocket launch data going back to 2005, I discovered that Israel's enemies were carrying out attacks in such a way that it was easy to predict when they would decide to increase the intensity of those attacks. By observing when Mars would be within 30 degrees of the lunar node within a calendar year (January thru December), I was able to find a strong correlation between the escalation of rocket fire from Gaza into Israel relative to the rest of the year. It was discovered that since 2005, Gaza militants would fire their highest intensity of rocket fire when Mars would be within 30 degrees of the lunar node. After years of successful predictions, it becomes warranted for me to provide a scientific explanation that would help elucidate this matter. First, let me provide the basis and justification for embarking upon an investigation into Mars influence on human behavior.

The Mars Effect, first introduced by French researcher Michel Gauquelin in 1955, is a thesis that provides statistical evidence supporting a connection between the position of the planet Mars and the eminence of sports champions. The evidence showed that there was statistical significance for Mars appearing in key areas of the astrological charts of major sports champions. Gauquelin divided the chart into 12 sectors and, in his research of astrological charts for thousands of elite athletes, discovered that Mars was positioned in the key sectors called the rising sector and the culminating sector with greater than chance probability. The base rate for a planet to appear in 2 out of 12 sectors, based on chance, was 17%. In Gauquelin's extensive data samples, Mars appeared with a frequency of 22%, which is more than coincidence and thus - putting aside all other possible meanings - means that Mars must have some effect. So this finding is enough to rationalize a belief in Mars influence.

Professor Suitbert Ertel came along in the 1980s and developed a criterion for calculating eminence by counting the number of citations of a particular athlete in sports reference books. The greater the number of citations, the greater the eminence. In his test, using Gauquelin's collection along with his own eminence criteria, he found that the Mars effect played a stronger role among athletes with higher citation counts, confirming Gauquelin's hypothesis that Mars appears more frequently in key sectors on the charts of major sports champions . The significance of Gauquelin's work is that it was the first time that astrology had ever been given scientific consideration. The work of Gauquelin and Ertel is the spark strong enough to justify belief in the influence of Mars and provides a strong foundation for the formation of a new system based on science and empirical data.

After Gauquelin and Ertel linked the influence of Mars to a scientific potential, I took Mars and linked it to a religious meaning. I had undertaken solving an age old mystery concerning the number of the beast, 666, which comes from the Christian biblical literature. 666 is a number that has garnered much suspense because it is a number associated with Satan, the great adversary and enemy of God and God's people. In Christian tradition, 666 is defined as the number of the beast and over the centuries, there have been many attempts to figure out what and who this number stands for. Traditionally, this number is associated with a person, but others have assigned to systems and kingdoms. In any event, there were myriad attempts by scholars and mystics alike to solve the mystery of 666. I embarked upon solving the riddle and came up with Mars 360, which is the revolution of Mars around the Sun and its influence on humanity.

Using English Sumerian gematria, in which the letters of the alphabet are numbered in multiples of 6...A=6, B=12, c=18, etc., I added up the letters of Mars and got 306. After simply adding 360 to 306, I came up with 666 and tied Satan to Mars influence or Mars 360. Keep in mind that, in Jewish Talmudic tradition, Samael is the king of demons and an archenemy of Israel and is ruled by Mars. So here, we have a religious tradition that predates and surmises a future scientific understanding of Mars influence.

Combing this religious assertion with the scientific backing of Gauquelin's work regarding Mars influence on eminent sports champions and looking to see if Mars could apply to other Earthly matters related to the Abrahamic ethos of 666/the beast/Satan, I was able to discover that Mars' position within 30 degrees of the lunar node coincided with the escalation of rocket fire from Gaza into Israel since 2005. It is important to note that the adversarial competitive qualities denoted from the inference in Gauquelin's research that Mars influences sports champions can also be applied to soldiers or terrorists in situations where the end game is dominance or destruction of an adversary. I discovered this correlation in 2019. After discovering this, I was able to prove this was the case in real time. In my research, the statistics show that Mars usually makes a complete transit within 30 degrees of the lunar node over a period of about 3 to 3.5 months each calendar year, unless Mars goes retrograde during the alignment which can prolong the time of that configuration. The base rate for predicting that something will happen within a roughly three-month period within a calendar year is around 30.0%. Essentially, anyone who randomly picks 3.5 months within a calendar year has about a 30% chance of predicting the time frame in which the largest rocket fire from Gaza into Israel would occur. However, between 2019 and 2024, using the observation of Mars, I was accurate in predicting when the highest concentration of rocket launches against Israel would occur at a 100% success rate. In 2020, Mars was within 30 degrees of the lunar node between January 15 and April 3. According to the data, that period covered the highest concentration of rocket launches against Israel compared to the entire year of 2020. About 115 rockets were launched at that time, more than any other time in 2020. In 2021, Mars underwent a full phase of being within 30 degrees of the lunar node between February 9 and May 13. Over 4,000 rockets were fired at Israel toward the end of that phase, more than any other time in 2021. In 2022, Mars went through a full phase of being within 30 degrees of the lunar node between June 22 and September 19. About 1,100 rockets were fired at Israel during that time period in early August, more than any other time in 2022. In 2023, Mars went through a full phase of being within 30 degrees of the lunar node between August 24 and November 15, and during that time, terrorists fired 10,000 rockets at Israel, more than any other time in 2023. In 2024, Mars went through a full phase of being within 30 degrees of the lunar node between April 12 and June 25. During that period, terrorists fired approximately 3,500 rockets, which already exceeded the amount fired at any other time in 2024.

According to the data on rocket fire from Gaza since 2005, Israel's enemies, including terrorists from Gaza, as well as terrorists from Lebanon and Iran, have fired a total of 37,166 rockets at Israel since 2005. Since 2005, 25,184 rockets have been fired at Israel while Mars was within 30 degrees of the lunar node. At any other time since 2005, 11,982 rockets have been fired at Israel. 68% of the total rockets fired at Israel since 2005 were fired while Mars was within 30 degrees of the lunar node. In the 15/20 years between 2005 and 2024, most of the rockets fired during the calendar year were fired while Mars was within 30 degrees of the lunar node. In the twenty years between 2005 and 2024, the month with the highest number of rockets fired at Israel during the year coincided with Mars being within 30 degrees of the lunar node. This is a 100% correlation.

Of course, after encountering many skeptics who would often apply the caveat that correlation doesn't equal causation, I have become obliged to provide a more biological and geological explanation that could elucidate this Mars thesis beyond just statistical analysis. But it must said that any undertaking that applies inductive reasoning must put forth a prediction model and here we already have one. In any case, lets go through examples of some of the theories put forth on how Mars or celestial bodies could have an effect on human behavior.

During Gauquelin's work on the Mars effect, there were numerous attempts to explain how Mars might exert a geological or biological influence on human behavior. Gauquelin suggested that the birth of the fetus was triggered by its response to planetary signals. Frank McGillion, author of The Opening Eye, explained this further by hypothesizing that the signals are sensed by the pineal gland. Jacques Halbronn and Serge Hutin, authors of Histoire de |'astrologie, later postulated that a person's beliefs are shaped genetically. In 1990, Percy Seymour, the author of The Evidence of Science, tried to explain that the signals emitted by planets are the result of the interaction between planetary tides and the magnetosphere. Peter Roberts assumed that the signals from planets are perceived by the human soul. German psychology professor Arno Miiller argued that men born with prominent planets were the dominant men with the most reproductive rights. Ertel tried to find out whether there was a physical basis for the Mars effect. He tested Mars in relation to Earth and tested whether the distance between Earth and Mars would cause variations in the Mars effect. Angular size, declination, orbital position relative to the Sun, and geomagnetic activity on Earth were all ruled out by Ertel as anything that could physically explain the Mars effect. I explain the Mars phenomenon further by positing and demonstrating how Mars creates an effect when it is within 30 degrees of the lunar node. The essence of this alignment and hypothesis is essentially that the closer the planet Mars is to the intersection between the Moon's orbit and Earth's orbit, an effect is created that causes people to exhibit more pessimistic, cynical, and aggressive characteristics place. During this phase, stock market investors are negative about the market while militants become more aggressive compared to other times when Mars is not within 30 degrees of the lunar node. The basic premise that is easily justifiable is that if the moon is exerting a gravitational tug on the tides of the ocean, and that since humans are made up of mostly water, then it is thus rational to believe that the moon can have an effect on the human behavior. However, I inferred that Mars must also exhibit a similar effect as the moon.

The lunar nodes are the intersecting points between the Moon's orbital plane around the Earth and the Earth's orbital plane around the Sun. Starting within 30 degrees of the lunar node, the closer Mars' orbit around the Sun gets to the intersection (lunar node) between the Moon's orbit around the Earth and the Earth's orbit around the Sun, the greater the influence of Mars on Earthly events. The best physical explanation I can give may have to be derived from the influence of the moon. I suggested that since the Moon has been confirmed to exert a gravitational pull on the Earth, such that the closer the Moon is to the Earth, the higher the ocean tides are, then the Moon must also influence people's moods because the human body consists mostly of water. Since this Mars explanation is based on its position in relation to the intersection between the Moon's orbital plane and the Earth's orbital plane, I assert that Mars could exert influence on humans in a similar manner. My big break came in 2024 when scientists discovered that Mars was exerting a strong gravitational pull on Earth, bringing Earth closer to the sun, leading to phases or warming and cooling that spans over 2 millions years. Keep in mind that my postulates about Mars, as well as Gauquelin's, predates this scientific finding that Mars does in fact have an effect on Earth. And here we see now in 2024, scientists are beginning to posit that Mars does have an effect on Earth's climate and ocean tides, which confirms my thesis, as well as Gauquelin's.

Here is an article from science.org "The Moon causes both high and low tides, but it's not the only celestial body that impacts Earth's waters. Mars's gravity influences our planet's deep-ocean currents, according to a study reported in Nature Communications this week."

Here is an excerpt from the article.

study reported in *Nature Communications* this week. By comparing more than 50 years of deep-sea drilling records with shifts in Earth's orbit, researchers found that the gravitational tug of Mars on Earth is causing it to wobble slightly on its axis. Every 2.4 million years, Mars's orbit comes close enough to Earth that its gravity can affect it, tilting Earth's usual path and orientation. This orbital shift causes Earth to be exposed to more sunlight, warming the climate, which, in turn, stirs up ocean currents and makes them stronger. However, some researchers doubt that Mars's weak gravitational pull is the true cause of these changes, *New Scientist* reports.

What this does is open the flood gates on Mars influence and with this information, we can develop more insight into how Mars influences human behavior. According to this scientific finding, as Mars travels around the sun, it exerts a gravitational pull on Earth, eventually affecting Earth's axial tilt and orbital plane, which brings about periods of warming and cooling over long periods of time, millions of years in fact. With this understanding, we can posit that even during a calendar year, as Mars revolves around the sun, it is still exerting some measure of gravitational pull and some degree of warming, albeit a very minuscule one. This explains the revolution of Mars around the sun, which allows us to also explain how the lunar node is factored into all of this.

According to NASA, the moon is getting further away from earth by a distance of 3 centimeters each year, due to the moon's orbit expanding. My reservations would surmise that Mars may be the catalyst driving this effect when Mars goes within 30 degrees of the lunar node. Let me explain.

There is a gravitational force of attraction between all objects in the universe. The gravitational pull of a mass not only affects the position and orientation of other masses and vice versa, it can also affect the orbital planes of other masses and vice versa. This is what is happening when Mars goes within 30 degrees of the lunar node —essentially the mass of Mars is exerting a gravitational pull on the moon's orbital plane around the earth. It does this via the lunar node.

The lunar node is simply the point when the moon's orbital plane around the earth intersects the earth's orbital plane around the sun. That intersection point, I hypothesize, exposes the moon's orbital plane to Mars's gravitational pull when Mars is within 30 degrees of the lunar node, which in effect would, over time, bring the moon's orbital plane closer to the sun, which henceforth brings the moon itself further away from the earth at a distance of 3 cm each year. With the new understanding that as Mars orbits the sun and exerts a gravitational pull on the earth's axial tilt, leading to periods of warming and cooling over millions of years and even within smaller time-frames, we can now infer that when Mars goes within 30 degrees of the lunar node, Mars also exerts a gravitational pull on the moon's orbital path and stretches the orbital plane of the moon, bringing the moon further away from earth, which consequently would have a destabilizing effect on earth's wobble since it is the moon that is responsible for keeping the earth's wobble stable. Researchers ascertain that as the moon continues to get further away from earth, earth would consequently become subject to wild fluctuations in climate patterns since the moon's declining influence in stabilizing the earth's wobble would cause earth's wobble to become erratic, leading to drastic seasonal changes. By factoring in Mars, we can now make sense of this dynamic.

With this perspective, we can simply apply the corresponding aggression extrapolated from Mars influence to warmer temperatures, as there is a large body of scientific evidence linking aggression to higher temperatures—we can establish this as an axiom for our research into Mars influence on human behavior. But in this case, we should hypothesize that the corresponding aggression comes from higher temperature in relation to the mean and that these scenarios can be tied to Mars being within 30 degrees of the lunar node. Now, it can be inferred that when Mars is within 30 degrees of the lunar node, it is able to exert even more gravitational influence on earth's axial tilt by tugging on the moon's orbital plane, thereby widening the moon's orbital plane, incrementally bringing the moon further away from the earth and thus decreasing the moon's stabilizing influence on the earth's wobble, which would expose the earth to wilder fluctuations in temperature, even as Mars continues to exert a gravitational pull on the earth as it travels around the sun. This should thus affect temperatures and human behavior more severely. This could explain why there is evidence that human action is more drastic when Mars is within 30 degrees of the lunar node.

Those detractors of Mars influence can no longer dismiss Mars influence and relegate aggression from Gaza or the Middle East or anywhere to warmer temperatures that come about in the spring and summer. I can debunk the claim that militant aggression can simply be relegated to seasonal weather changes and not Mars influence.

Those stating that anyone can predict the highest escalation of rocket fire into Israel by anticipating for it to happen during the warmer months can put their theory to the test. Their theory gives a 7 month window, which is much greater than the one that I have at 3.5 months. Here are my timeframes of anticipating escalated rocket fire using Mars within 30 degrees of the lunar node, which has been accurate every year

Jan 15th 2020 - Apr 3rd 2020 - highest escalation occurred in February

Feb 9th 2021 - May 13th 2021 - highest escalation occurred in May

June 22nd 2022 - Sept 19th 2022 - highest escalation occurred in August

Aug 24th 2023 - Nov 15th 2023 - highest escalation occurred in October

April 12th 2024 - June 25, 2024 - highest escalation so far has occurred in May

If, in the last 5 years, one tried to predict that the highest escalation of rocket fire into Israel relative to the rest of the year would occur during the Spring and Summer months between March 20th and September 20th(a 7 month window), they would have been correct in 3 of the last 5 years. However, they would have been wrong in 2020 and in 2023, when it would have really counted, especially considering the scale of the attacks on October 7th 2023. So even with a 7 month window, one still would have failed to keep up with Mars within 30 degrees of the lunar node's 3.5 month window.

Furthermore, I can assert that warmer temperatures relative to the mean can bring about violence in the Middle East due to the gravitational pull of Mars on Earth by bringing it closer to the sun. It is easy to get confused here because one can note that

Mars is further away from the sun than the Earth is, thus leading one to believe that any gravitational pull from Mars would only bring Earth away from the sun. Visualizing how Mars and Earth orbit the sun, along with how there are times when Mars is closest to Earth and furthest away from Earth can help avoid confusion. As Mars orbits the sun, the further away it gets from the Earth, the more its gravitational pull brings the Earth's axial tilt closer to the sun. And in contrast, the closer Mars is to the Earth in its revolution around the sun, the more Mars's gravitational pull would bring Earth's axial tilt away from the sun. Here is a visual

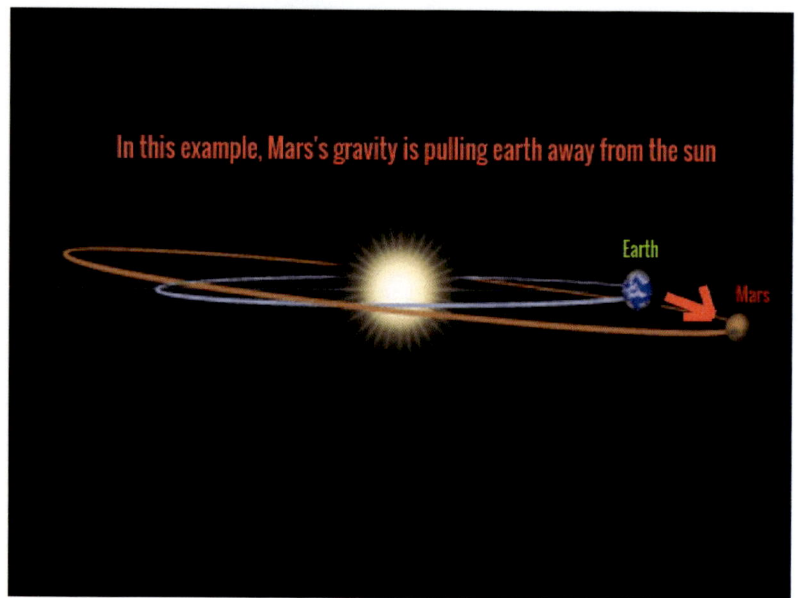

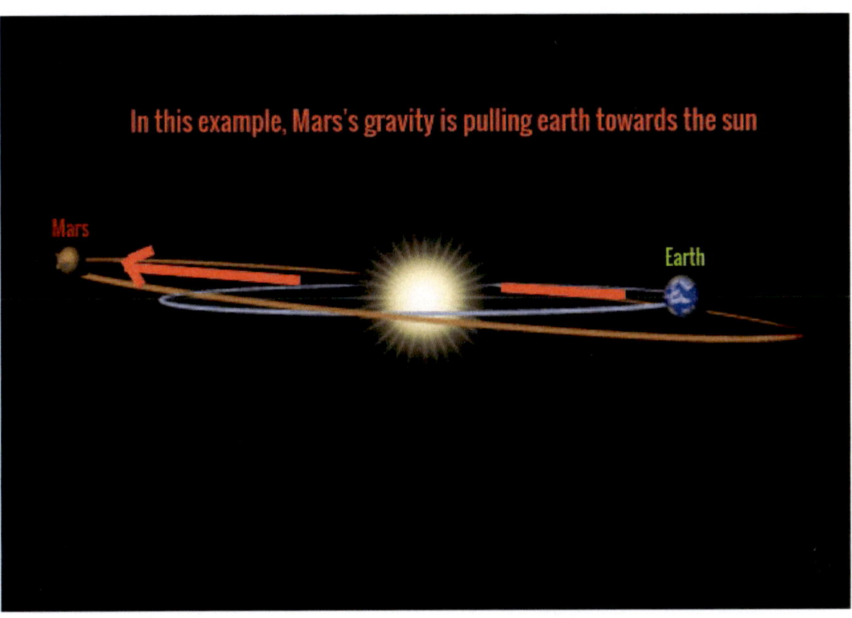

In this example, Mars's gravity is pulling earth towards the sun
Mars
Earth

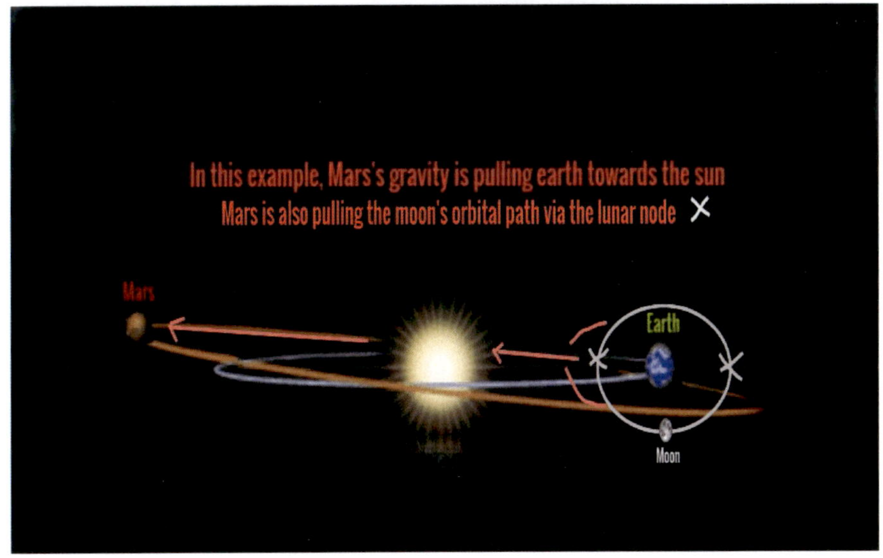

In this example, Mars's gravity is pulling earth towards the sun
Mars is also pulling the moon's orbital path via the lunar node X
Mars
Earth
Moon

With the the scientific discovery about Mars gravitational influence on Earth and the impact it has on Earth's climate, we can concur that this Mars effect of facilitating long periods of Earthly cooling and warming is the result of Mars slowly altering Earth's axial tilt and orbital path. During the time of Mars's gravitational influence, by which it pulls the Earth's tilt closer to the sun, exposing it to more solar radiation, the orbital plane of Earth also becomes affected, becoming more elliptical as time goes on, which exposes the Earth to more thermal radiation at perihelion than at aphelion. At the moment, the orbital path of Earth is close to circular with only a 6% difference in thermal radiation at perihelion than at aphelion.

Earth's tilt is the main factor that explains temperature changes, as opposed to Earth's proximity to the sun. In fact, in January, Earth is positioned closest to the sun, but during that time temperatures are cooler. While in July, the Earth furthest away from the sun, but yet temperatures are warmer. The reason for this dynamic is explained in how Earth's axial tilt affects how sun rays hit the Earth. During the summer the rays from the sun hits the Earth at a steep angle, and does not spread out, which leads to a greater concentration of energy hitting the Earth. This is in contrast to the winter, when the sun hits the Earth at a more shallow angle, where the sun rays are more spread out and less intense in terms of energy. One can try to apply this dynamic to the Gaza rocket fire situation, but as already explained, using Spring and Summer months would have led to miscalculation in 2 of the 5 years I used as an example. If we factor Mars in, we can began to posit that the location of Mars in relation to the Earth will affect the mean temperatures in any given season. For example, lets say Mars is furthest away from the Earth, within 30 degrees of the lunar node, but exerting a gravitational tug on Earth's axial tilt, bring the angle closer to the sun, albeit to a minute degree. The result, theoretically, regardless of the season, should be a higher mean temperature, perhaps more precipitation, and thus expose human beings to a higher level of aggression. Here is an example. Here is a visual of how Mars lined up with Earth on October 7th, the day that Hamas launched a massive terror operation against Israel. Mars was within 30 degrees of the lunar node, but far away from Earth, but exerting gravitational forces pulling Earth's axial tilt towards the sun

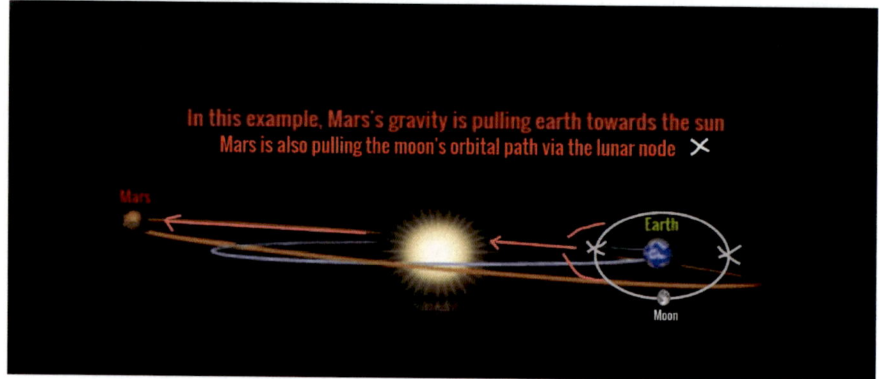

Here is the astrology chart for October 7th. In an astrology chart, Earth is always opposite the sun. I added an icon

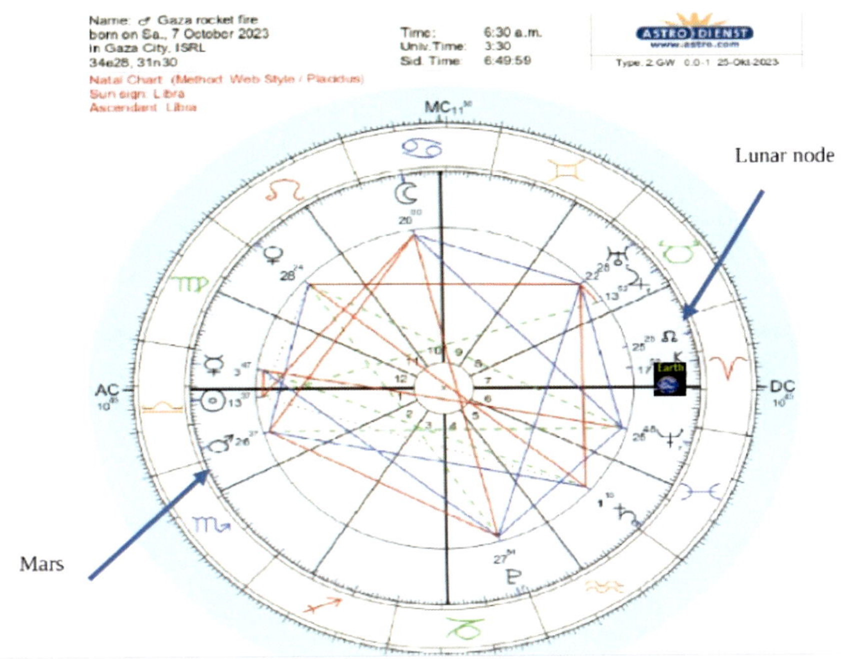

This is an example of a major attack happening in the fall, not a typical timeframe known for compelling aggression. So thus we can look at the Mars factor here. I have posited that Mars influence on aggression relates not to higher temperatures in general, but to higher temperatures relative to the average. October 2023 was the warmest October ever recorded.

The planet just had its warmest October on record

So far, 2023 is a record-warm year for the globe

November 15, 2023

October 2023

The average global temperature for October was 2.41 degrees F (1.34 degrees C) above the 20th-century average of 57.1 degrees F (14.0 degrees C), ranking as the world's warmest October on record. This was 0.43 of a degree F (0.24 of a degree C) above the previous record from October 2015. For the seventh-consecutive month, global ocean surface temperature also set a record high.

There is a large corpus of information, studies and research that link higher temperatures to aggression and lower cognitive function, however, in regards to the Mars thesis and how it influences aggression, I posit that higher temperatures in relation to the mean is what triggers aggression and lowers cognitive function. I also infer that these higher temperatures in relation to the mean should theoretically bring above average rainfall.

I have appended the Gaza rocket fire data to this document for reference.

Please note that the data from 2006 includes rocket fire from Gaza and Lebanon. The 2006 Israel - Lebanon war began on July 12th 2006 and ended on August 15th 2006

The data from 2024 includes rocket fire from Gaza, Lebanon and Iran

There is a pattern in which the highest concentration of rocket fire against Israel within a calendar year is occurring when Mars is within 30 degrees of the lunar node. This is happening at a rate of 70% since 2005.

In every single year since 2005, the month containing the highest rocket fire for the year is a month in which Mars had been within 30 degrees of the lunar node at some point.

Figure H - Gaza Rocket attacks on Israel
* = Largest amount of Rocket fire for the year

	2005	2006	2007	2008	2009	2010	2011	2012
Jan	40	7	28	136	345*	13	17	9
Feb	5	9	43	228	52	5	6	36
Mar	23	41	31	103	34	35*	38	173
Apr	34	79	25	373*	5	5	87	10
May	77	54	257*	206	1	14	1	3
Jun	129	140	63	153	2	14	4	83
Jul	211*	2135*	61	4	1	13	20	18
Aug	50	2087	81	8	1	14	145*	21
Sept	61	40	70	1	-10	16	8	17
Oct	26	52	53	1	1	3	52	116
Nov	42	157	65	125	4	5	11	1734*
Dec	76	50	113	361	4	15	30	1

cont'd
Figure H - Gaza Rocket attacks on Israel
* = Largest amount of Rocket fire for the year

	2013	2014	2015	2016	2017	2018	2019
Jan	0	22	0	6*	0	6	0
Feb	1	9	0	0	7*	4	0
Mar	4	65	0	5	2	0	3
Apr	17*	19	1	0	1	0	0
May	1	4	1	2	1	70	600*
Jun	5	62	3	0	1	64	3
Jul	5	2,874*	1	2	2	174*	0
Aug	4	950	3	1	1	8	0
Sept	8	0	4	0	0	0	1
Oct	3	1	5*	0	1	0	0
Nov	0	0	3	0	0	17	455
Dec	4	1	4	0	7	0	4

Sources for rocket fire data come from
The Jewish Virtual Library, Wikipedia, and the Israeli Army Radio
https://www.shabak.gov.il/reports/

cont'd
Gaza Rocket attacks on Israel
* = Largest amount of Rocket fire for the year

	2020	2021	2022	2023	2024	2025	2026
Jan	8	3	0	1	691		
Feb	104*	0	0	8	699		
Mar	0	0	0	0	850		
Apr	0	45	5	66	1037		
May	1	4375*	0	1470	1452*		
Jun	3	0	1	0	1060		
Jul	3	0	4	6	1307		
Aug	15	1	1100*	0			
Sept	13	2	0	0			
Oct	3	0	0	8500*			
Nov	3	0	4	2000			
Dec	2	0	1	1000			

Dates of Mars-within-30-degrees-of-the-lunar node

May 29 2005 - Aug 29 2005   **Nov 17 2005 - Dec 27 2005**   **Jul 20 2006 – Oct 14, 2006**

March 19 2007 - May 30 2007   Apr 28 2008 - Jul 31 2008   Jan 08 2009 - Mar 24 2009   Aug 24 2009 - May 02 2010
Nov 02 2010 - Jan 18 2011   Jun 11 2011-Sep 01 2011   Aug 24 2012 - Nov 12 2012   Apr 03 2013 - Jun 22 2013
Dec 19 2013 - Aug 28 2014   Jan 27 2015 - Apr 12 2015   Sep 27 2015 - Dec 26 2015   Nov 21 2016 - Feb 01 2017
Jul 11 2017- Oct 10 2017   Apr 08 2018 - Nov 14 2018   May 01 2019-Jul 29 2019   Jan 15, 2020 - Apr 3rd 2020

> **Below are future dates of Mars within 30 degrees of the lunar node**

Feb 9, 2021 - May 13, 2021
Nov 4, 2021 - Jan 22, 2022
June 22, 2022 - Sept 19, 2022
Dec 26 2022 - Jan 24, 2023

Aug 24, 2023 - Nov 15, 2023
April 12, 2024 - June 25, 2024
June 5, 2025 - Sept 4, 2025
Feb 4, 2026 - April 19, 2026
Sept. 27, 2026 - June 12, 2027

For five consecutive years, I have been able to predict when the highest concentration of rocket fire against Israel would occur within a calendar year.

In the last five years it was predicted that the highest escalation of rocket fire within the calendar year would occur during the time when Mars would be within 30 degrees of the lunar node.

1. Jan 15th 2020 - Apr 3rd 2020 - https://www.youtube.com/watch?v=e5GxO4ZW2fc
2. Feb 9th 2021 - May 13th 2021 - https://www.youtube.com/watch?v=v1sA-ZS73Lw&t
3. June 22nd 2022 - Sept 19th 2022 - https://www.youtube.com/watch?v=6EniwV0TWew&t
4. Aug 24th 2023 - Nov 15th 2023 - https://www.youtube.com/watch?v=IGbNPEO9qS4&t
5. April 12th 2024 - June 25, 2024 - https://www.youtube.com/watch?v=qW_-CiWu5b0&t

https://www.youtube.com/@anthonym1690

According to the Gaza rocket fire data going back to 2005 the enemies of Israel which include terrorists from Gaza, as well as terrorists from Lebanon and Iran have fired a total of 37,166 rockets at Israel since 2005 .

Since 2005 , 25,184 rockets were fired at Israel while Mars was within 30 degrees of the lunar node.

At any other time since 2005 ,11,982 rockets were fired at Israel

68% of total rockets fired at Israel since 2005 were fired while Mars was within 30 degrees of the lunar node

In 15/20 years between 2005 and 2024, most rockets fired during the calendar year were fired while Mars was within 30 degrees of the lunar node.

In 20/20 years between 2005 and 2024, the month containing the highest rocket fire for the year was also when Mars was within 30 degrees of the lunar node

Here are graphs of representing rocket attacks against Israel since 2005

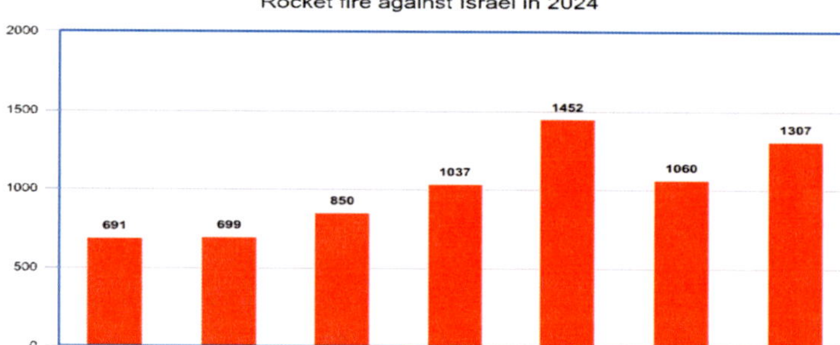

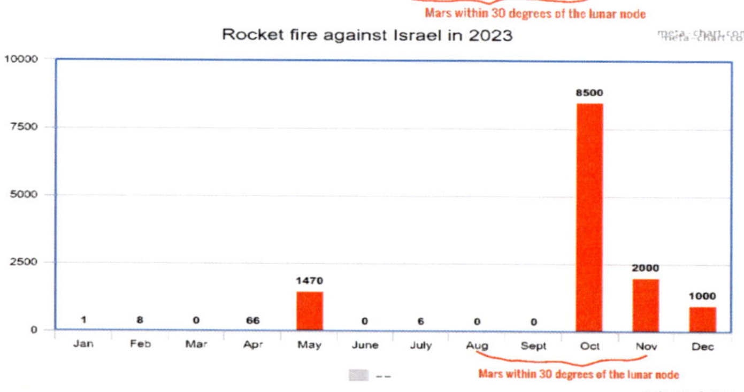

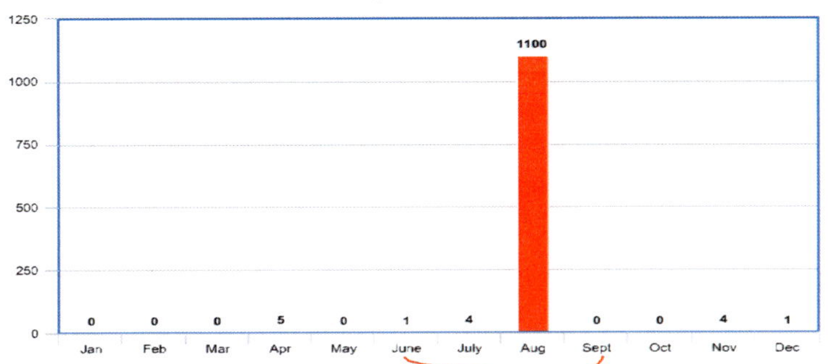

Rocket fire against Israel in 2022
1250
1000
750
500
250
0
1100
0 0 0 5 0 1 4 0 0 4 1
Jan Feb Mar Apr May June July Aug Sept Oct Nov Dec
-- Mars within 30 degrees of the lunar node
meta-chart.com

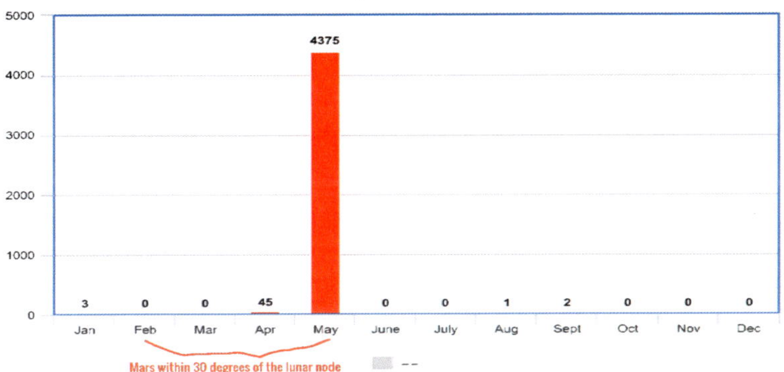

Rocket fire against Israel in 2021
5000
4000
3000
2000
1000
0
4375
3 0 0 45 0 0 1 2 0 0 0
Jan Feb Mar Apr May June July Aug Sept Oct Nov Dec
Mars within 30 degrees of the lunar node --
meta-chart.com

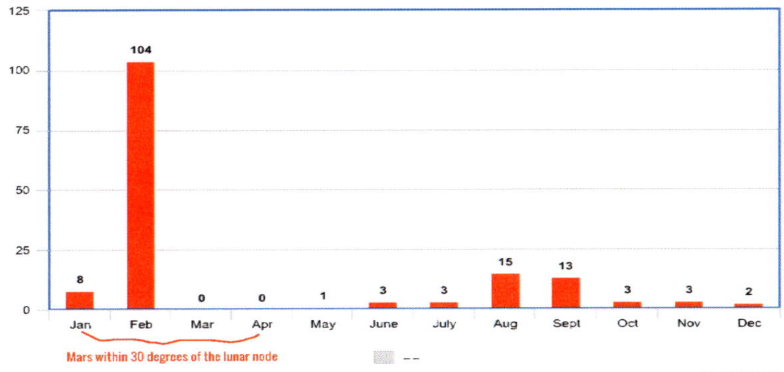

Rocket fire against Israel in 2020
125
100
75
50
25
0
104
8 104 0 0 1 3 3 15 13 3 3 2
Jan Feb Mar Apr May June July Aug Sept Oct Nov Dec
Mars within 30 degrees of the lunar node --
meta-chart.com

Rocket fire against Israel in 2019

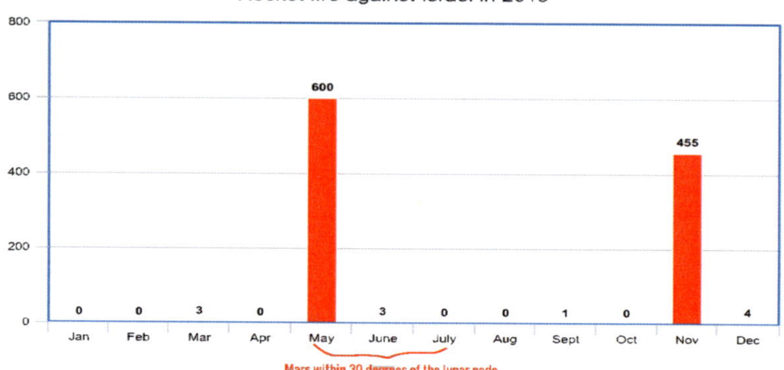

Rocket fire against Israel in 2018

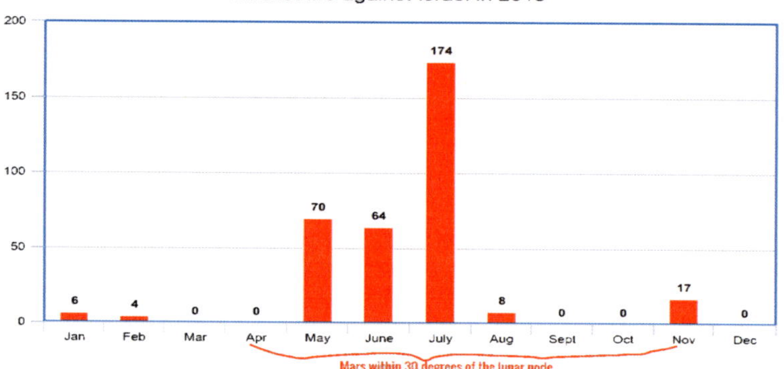

Rocket fire against Israel in 2017

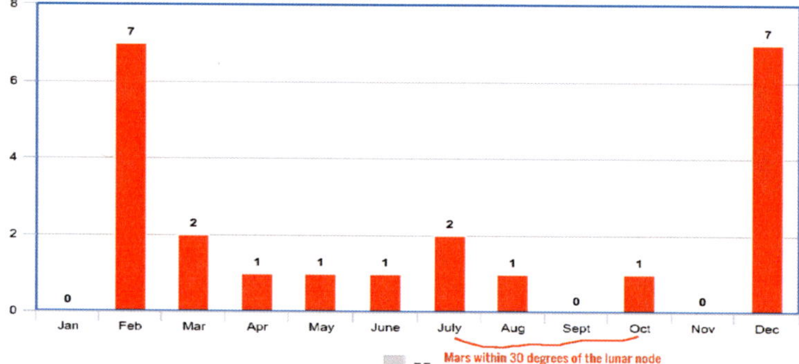

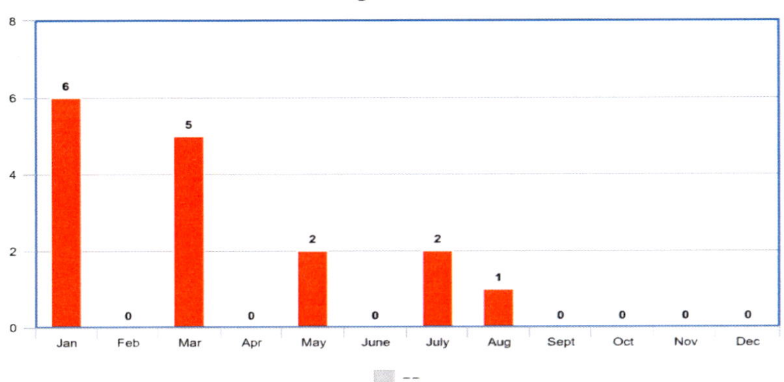

Rocket fire against Israel in 2016

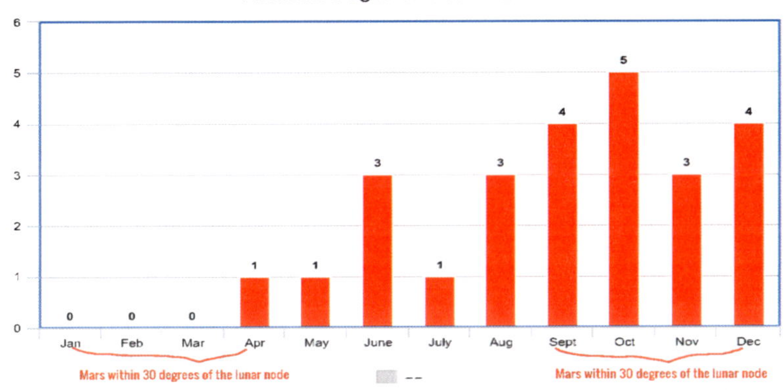

Rocket fire against Israel in 2015

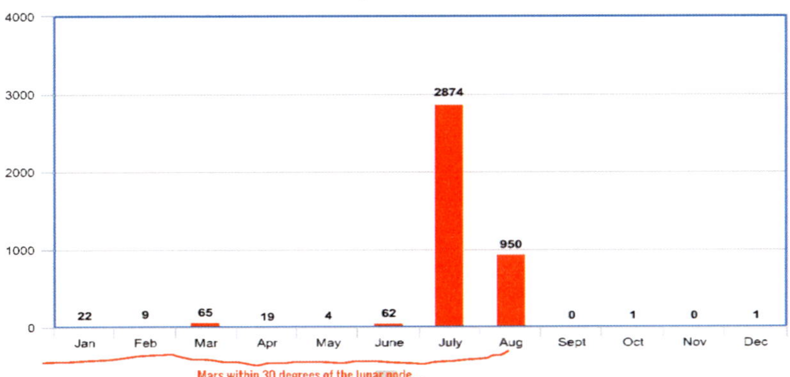

Rocket fire against Israel in 2014

17

Rocket fire against Israel in 2013

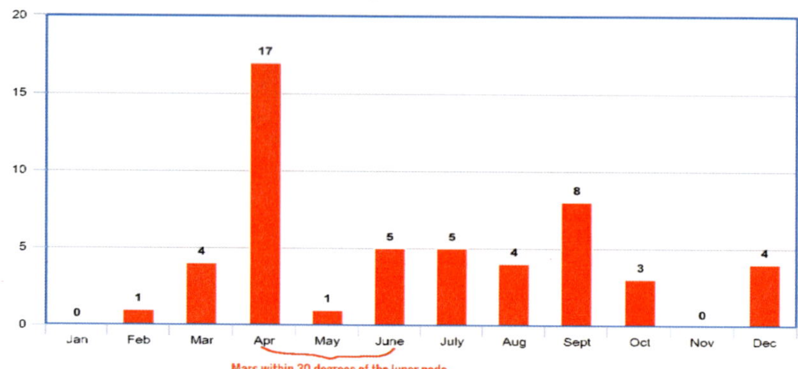

Rocket fire against Israel in 2012

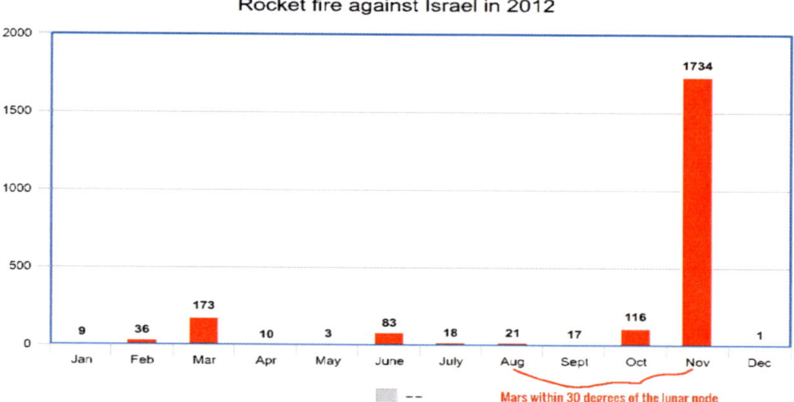

Rocket fire against Israel in 2011

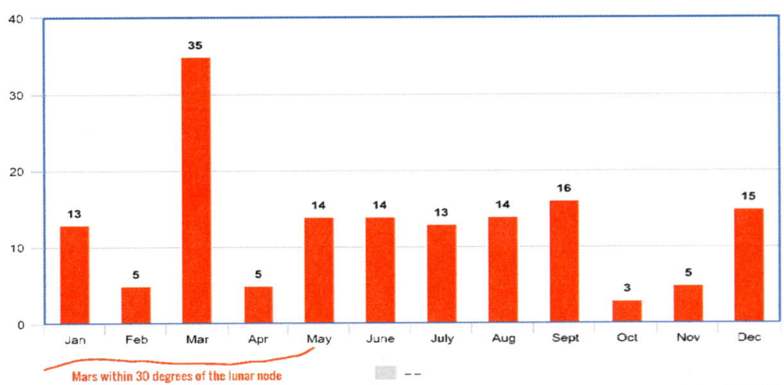

Rocket fire against Israel in 2010
40
30
20
10
0
13
5
35
5
14
14
13
14
16
3
5
15
Jan
Feb
Mar
Apr
May
June
July
Aug
Sept
Oct
Nov
Dec
Mars within 30 degrees of the lunar node
--
meta-chart.com

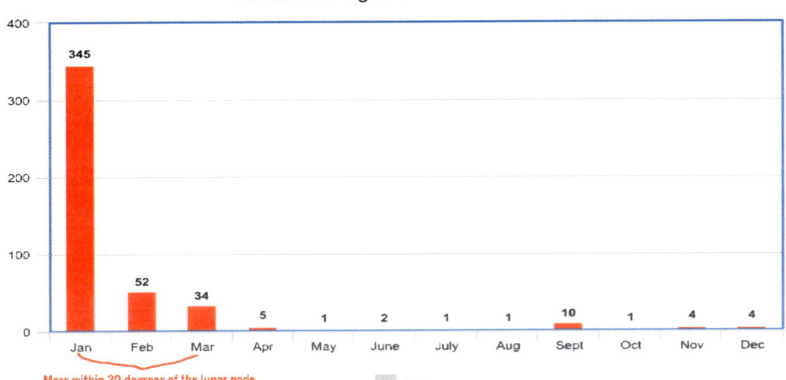

Rocket fire against Israel in 2009
400
300
200
100
0
345
52
34
5
1
2
1
1
10
1
4
4
Jan
Feb
Mar
Apr
May
June
July
Aug
Sept
Oct
Nov
Dec
Mars within 30 degrees of the lunar node
--
meta-chart.com

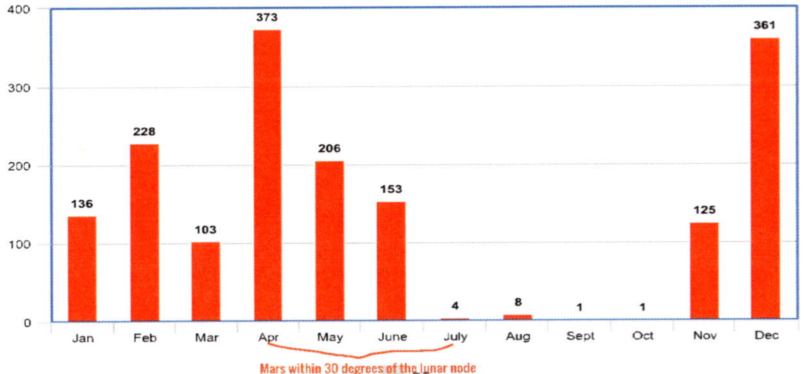

Rocket fire against Israel in 2008
400
300
200
100
0
136
228
103
373
206
153
4
8
1
1
125
361
Jan
Feb
Mar
Apr
May
June
July
Aug
Sept
Oct
Nov
Dec
Mars within 30 degrees of the lunar node
meta-chart.com

Rocket fire against Israel in 2007

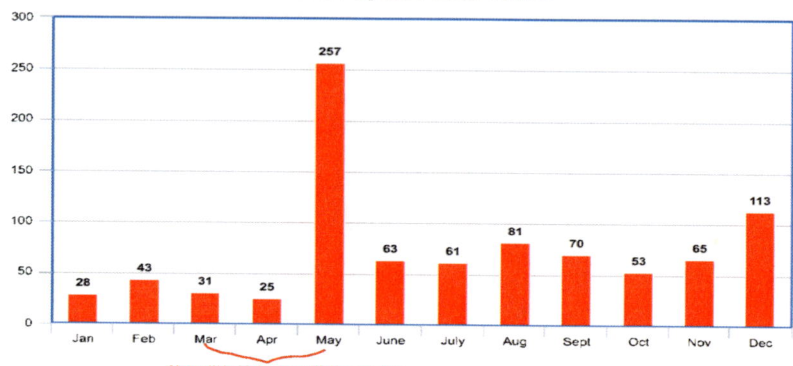

Rocket fire against Israel in 2006

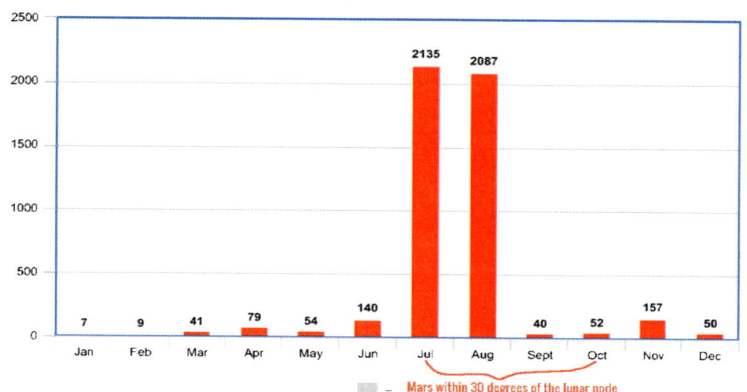

Rocket fire against Israel in 2005

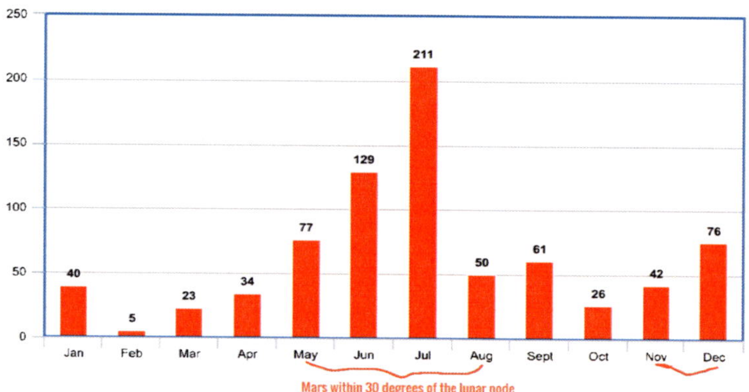

This section lays out the 25 major stock market crashes and downturns in US history. The data shows a 100% correlation between such events and Mars position in relation to earth. Every stock market crash and major stock downturn in US history has happened when Mars was orbiting behind the sun from earth's point of view.

To gain relevant context in regards to what this paper is demonstrating, it is important to take into account a recent study published in Nature Communications in March of 2024, roughly 5 years after this idea was first introduced to the public. In that study published in March of 2024, researchers discovered that Mars is exerting a gravitation pull on earth's tilt, exposing earth to warmer temperatures and more sunlight, all within a 2.4 million year cycle. I assert that this allows us to surmise that, even within smaller timeframes, Mars is still exerting a gravitational pull on earth's axial tilt, enough to raise temperatures when the planet is within 30 degrees of the lunar node, which would affect human behavior. Citing the fact of numerous studies that link aggression and irritability to warmer temperatures, I establish an axiom and then assert that Mars within 30 degrees of the lunar node should affect the brain by reducing cognition and compelling aggression and irritability.

Here is a visual of what is happening as Mars travels around the sun and exerts a gravitational pull on Earth axial tilt. In this first graphic, Mars gravity is pulling earth's tilt away from the sun.

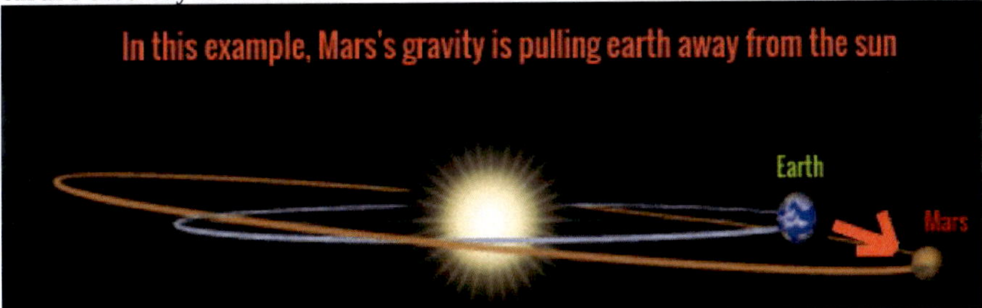

In this next graphic, Mars's gravity is pulling Earth's tilt toward the sun.

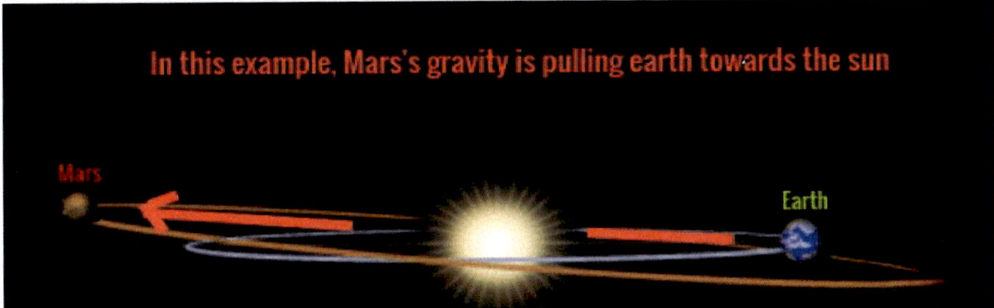

In this latter visual, this scenario should have the most prominent effect on human behavior. Here is how this scenario of Mars pulling earth's tilt towards the sun appears in an astrology chart. This is the chart for the October 29, 1929 Stock market crash. Planet earth is always opposite the sun in an astrology chart.

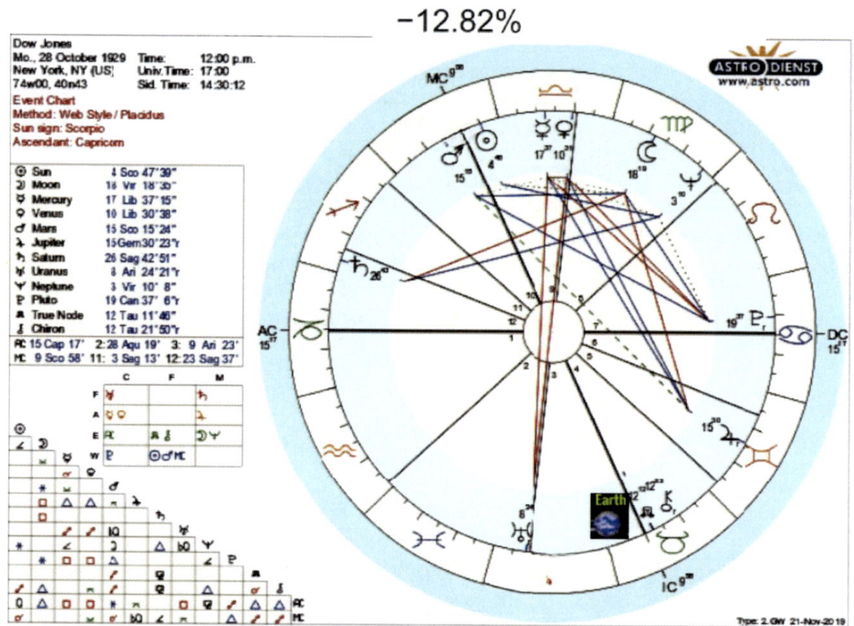

Here is the view from above of the same planetary configuration scenario.

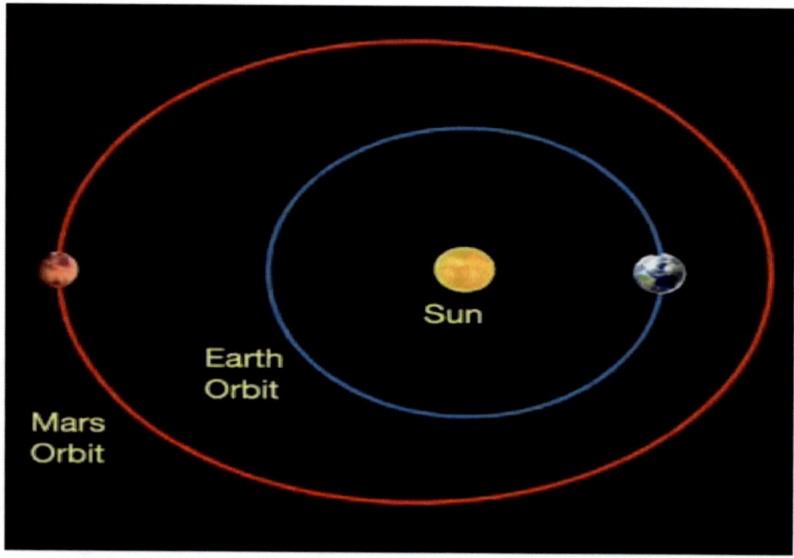

In all the major stock market crashes and one-day drops, Mars was somewhere along the white line as shown in this graphic, which according to the research, would indicate that Mars is pulling the earth's tilt towards the sun, triggering irritability

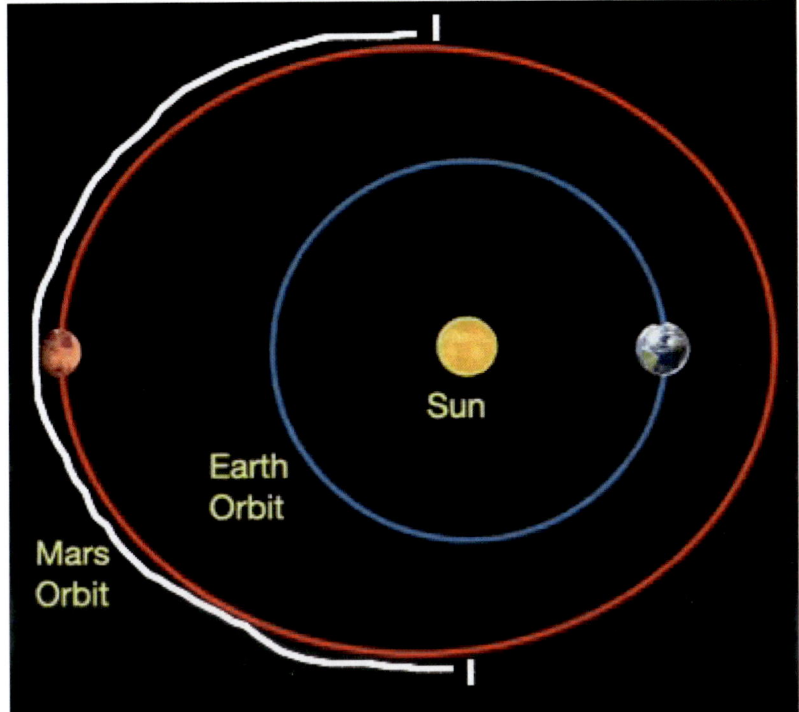

Here is how this same visual is represented in an astrology chart. See the next page Notice the white line.

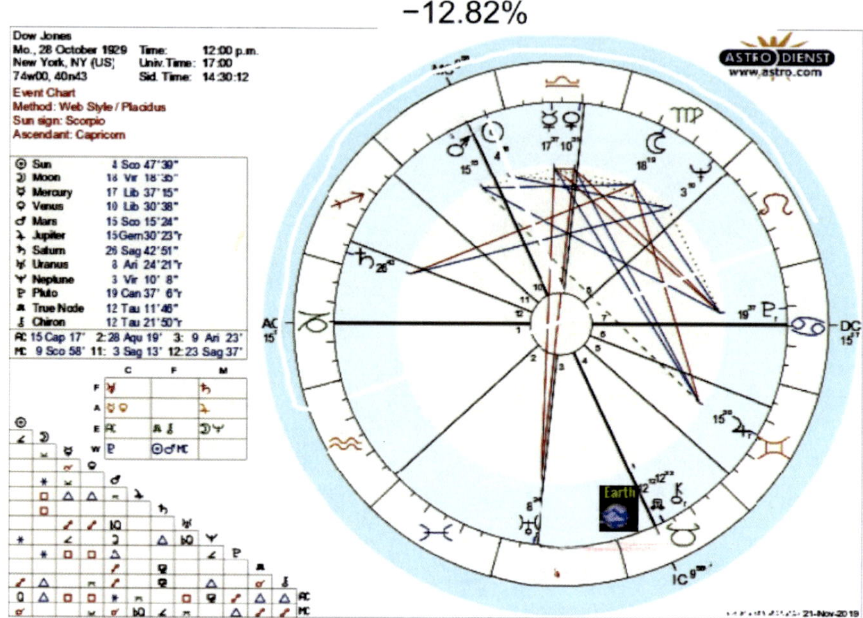

All the highest one day stock market crashes or one-day drops occurred when Mars appeared within that white line, as drawn out from the degree of the sun.

This perspective should help the reader move beyond the preconceived notion of absurdity and realize that this has scientific merit

Here are the rest of the stock market crashes, along with the astrology charts and the representation of Mars position in space

October 19, 1987 Stock market crash

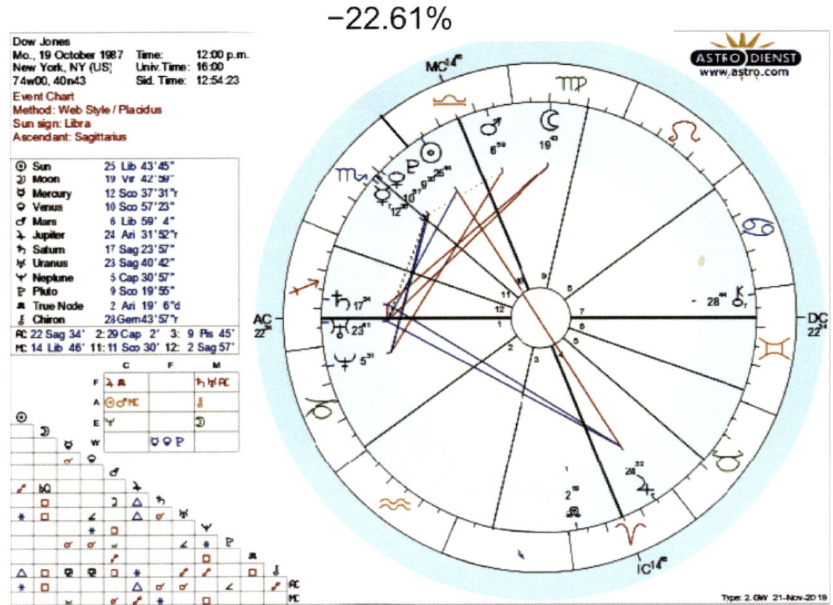

Here is how Mars was positioned in space that day relative to earth

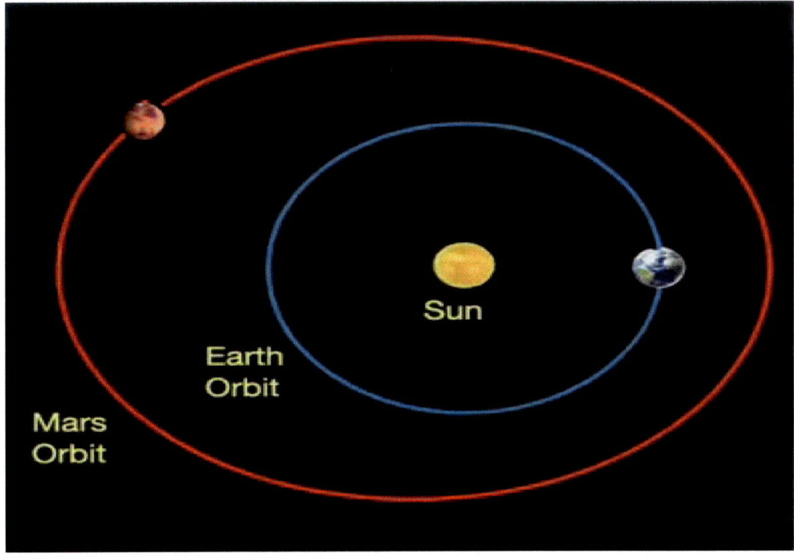

November 6 1929 Stock Market Crash

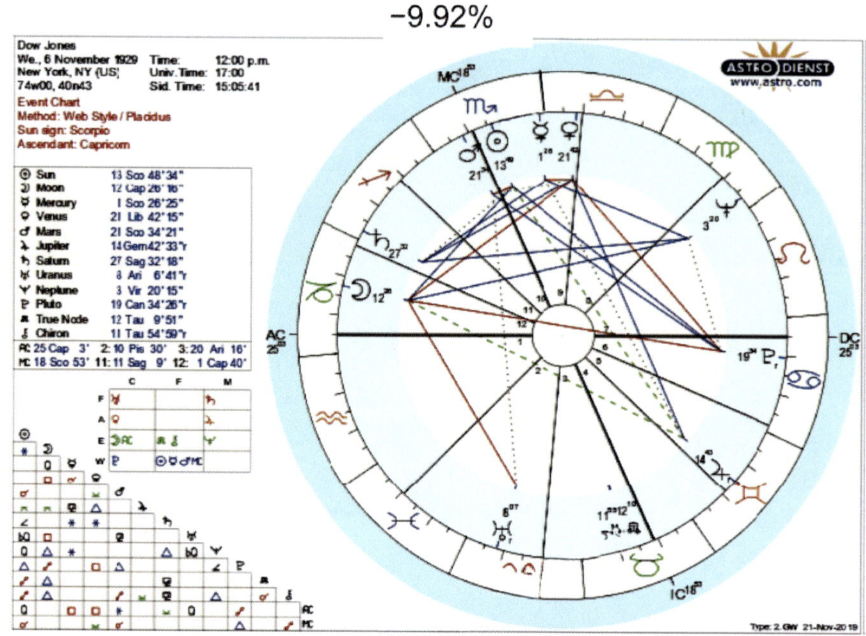

Here is where Mars was located in space from earth's point of view

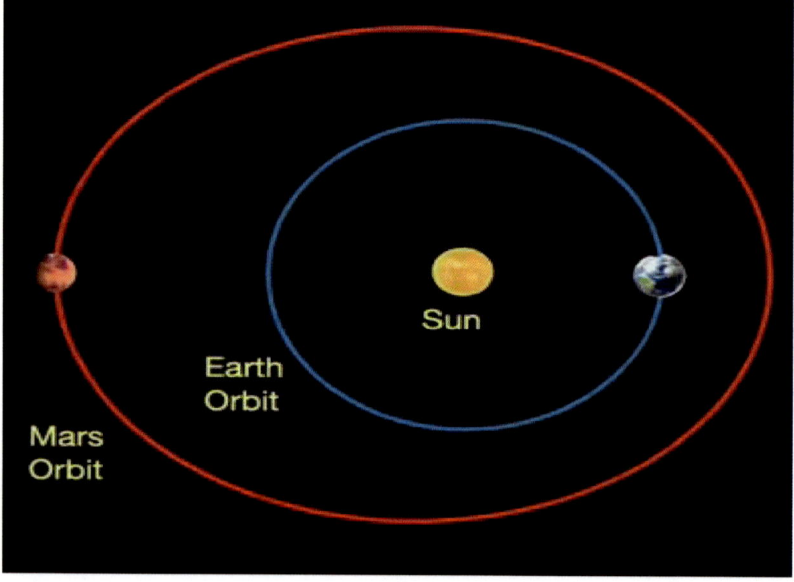

March 16, 2020 Stock Market crash

-12.93

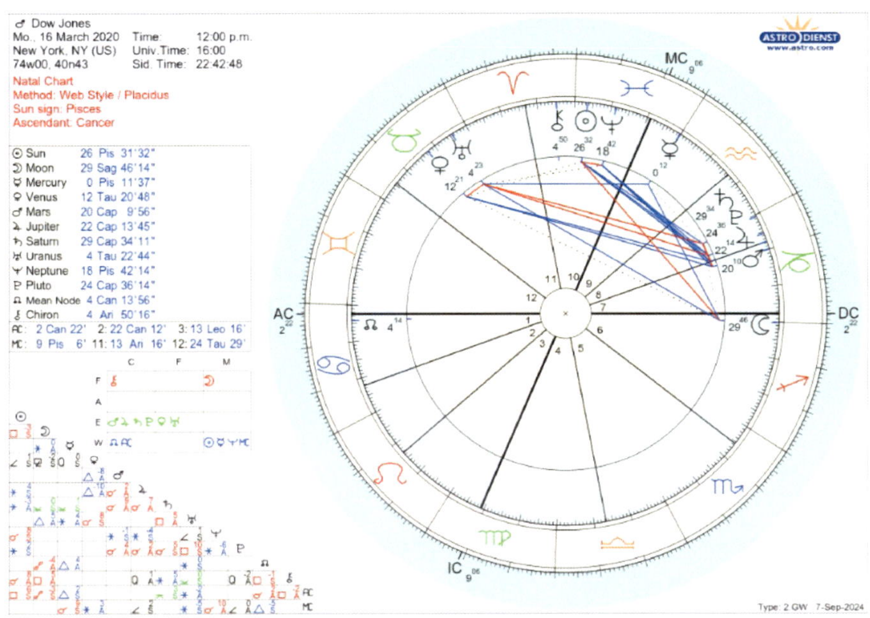

Here is how Mars was positioned in the sky relative to earth

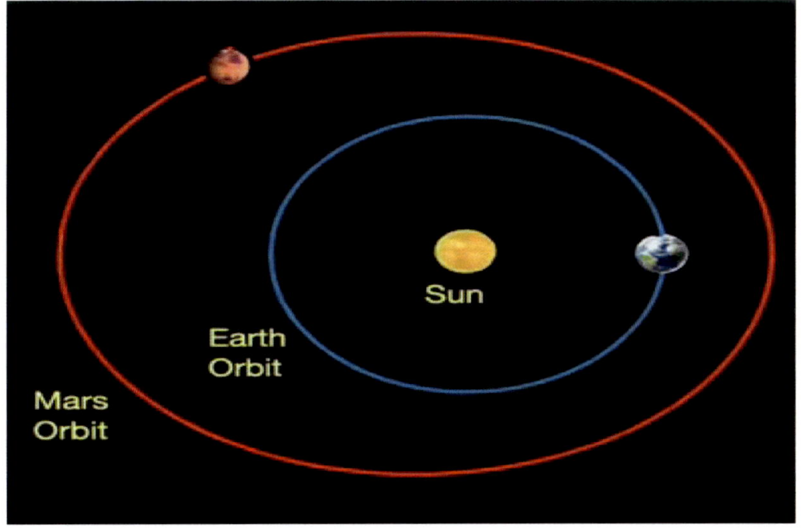

March 12, 2020 Stock Market Crash

-9.99

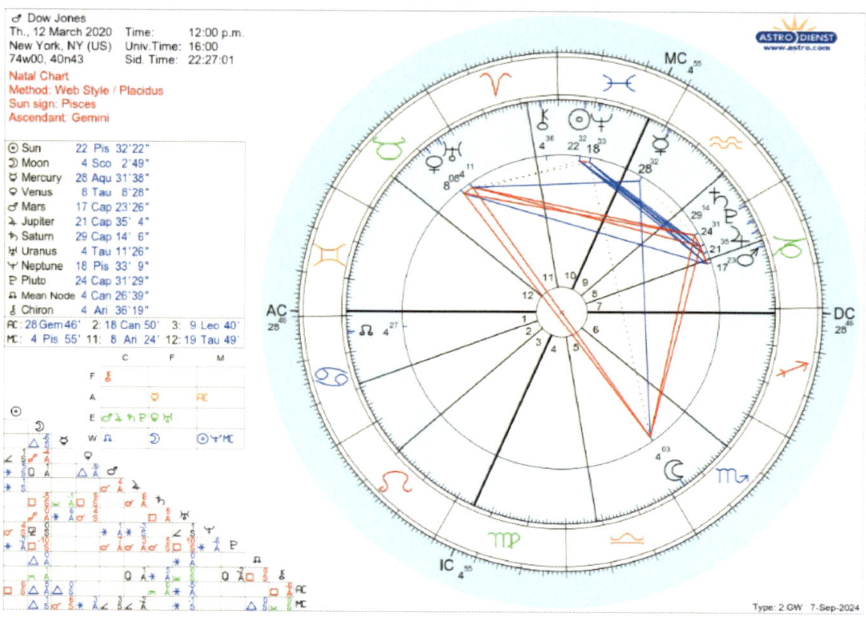

Here is how Mars was positioned in the sky on that day

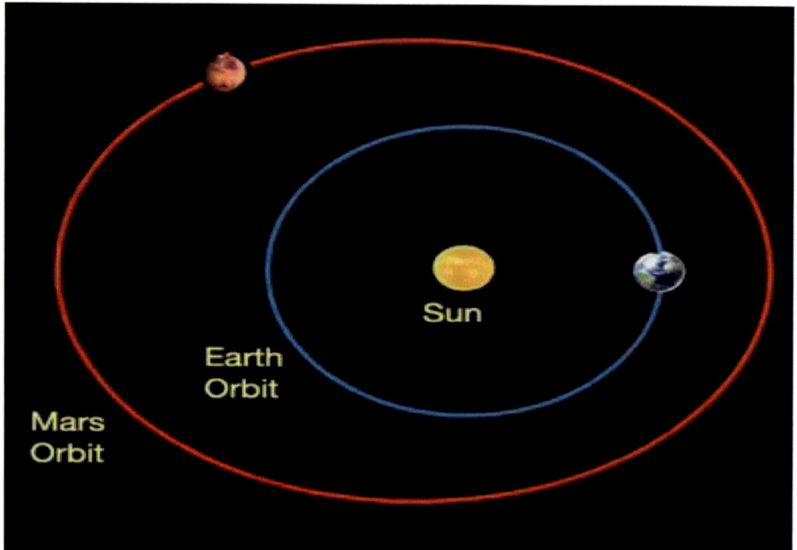

March 9, 2020 Stock Market Crash

-7.79

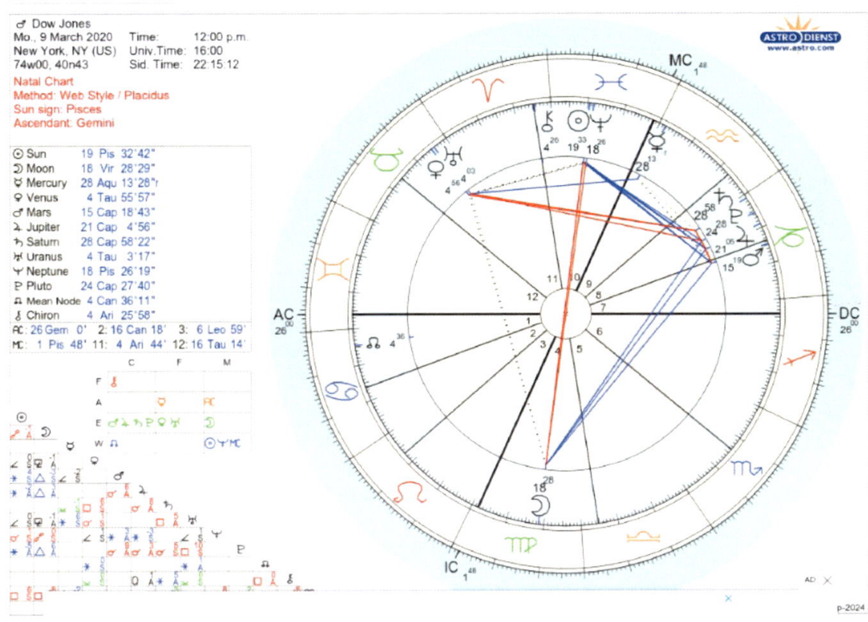

Here is where Mars was in the sky that day

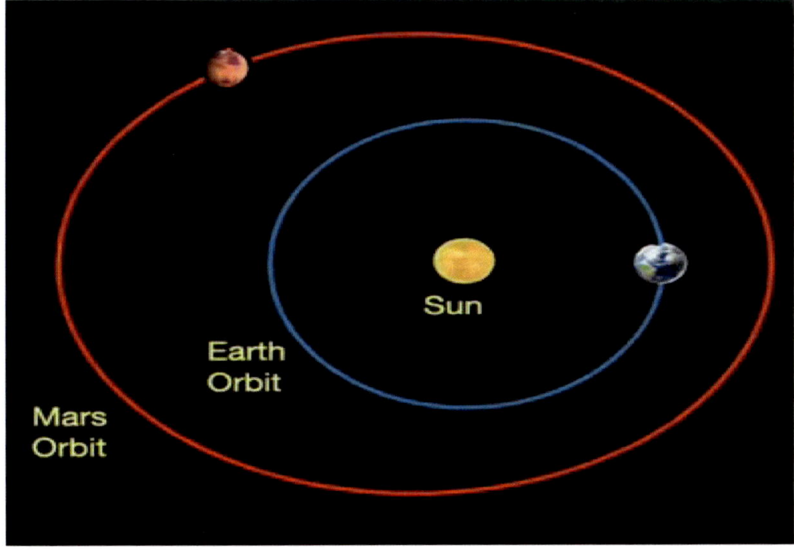

December 18, 1899, the stock market dropped -8.72 percent

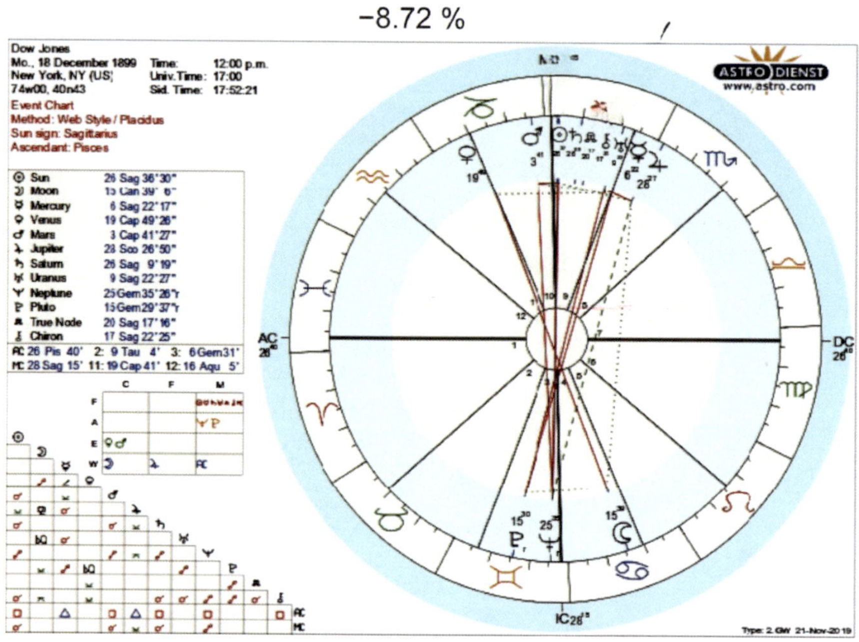

Here is how Mars lined up in the sky that day

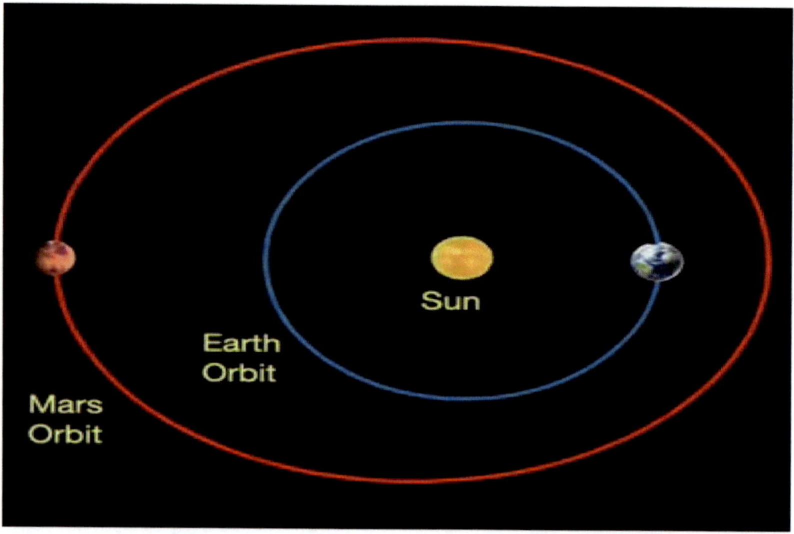

August 12, 1932, stock market dropped -8.4 %

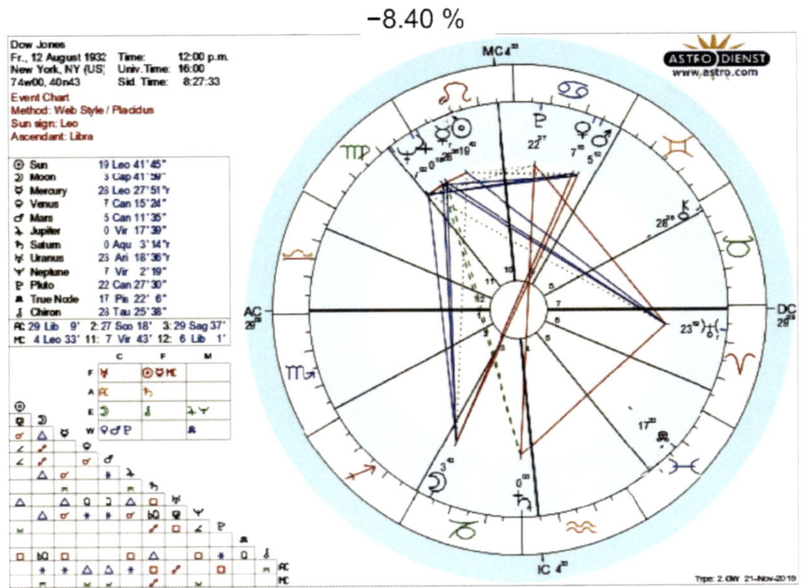

Here is how Mars was positioned in the sky that day

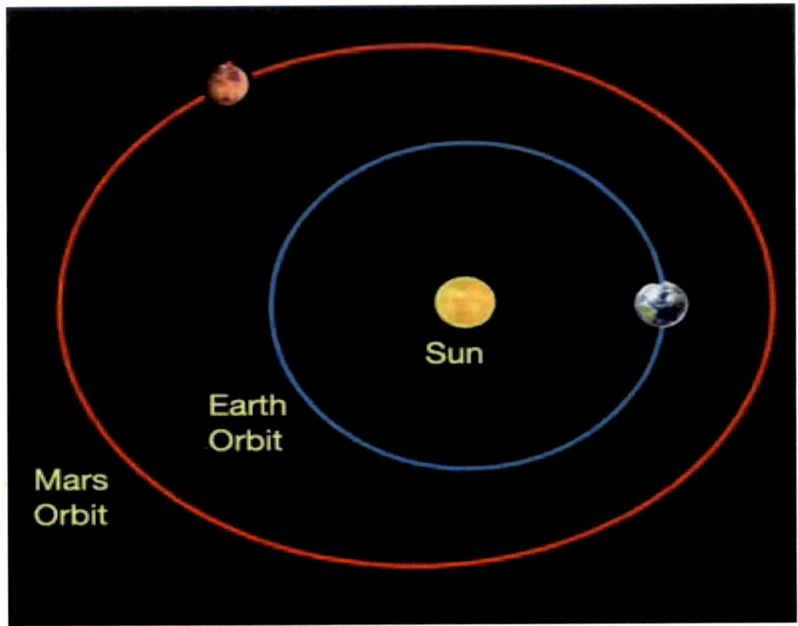

March 14, 1907, the stock market fell -8.29%

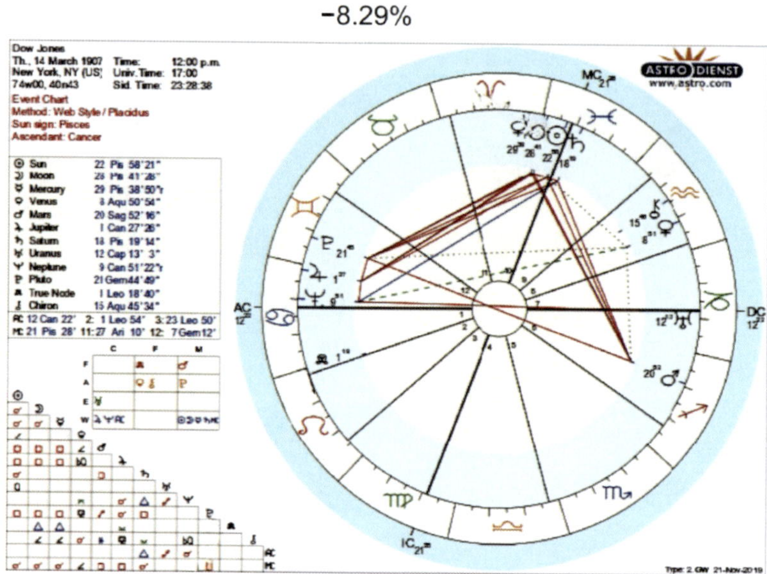

Here is how Mars was positioned in the sky that day

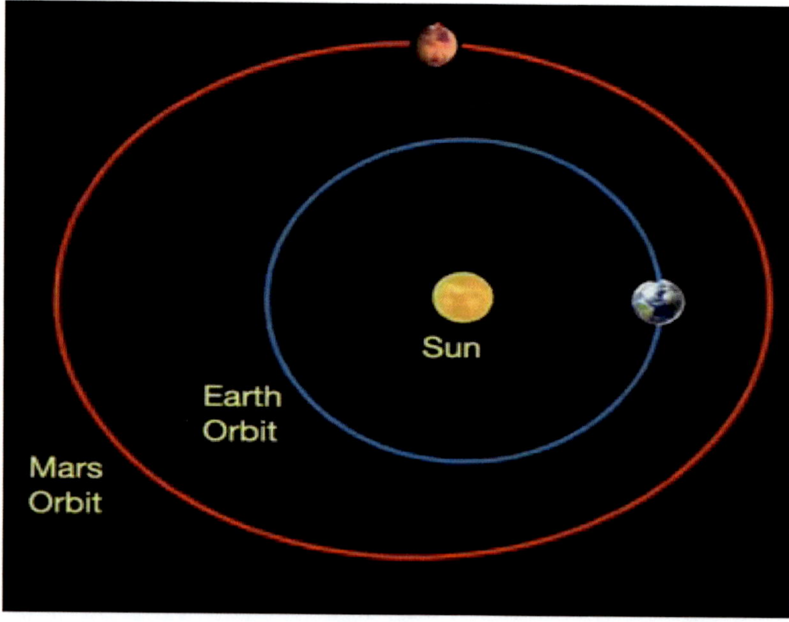

October 26, 1987, the stock market droped -8.04%

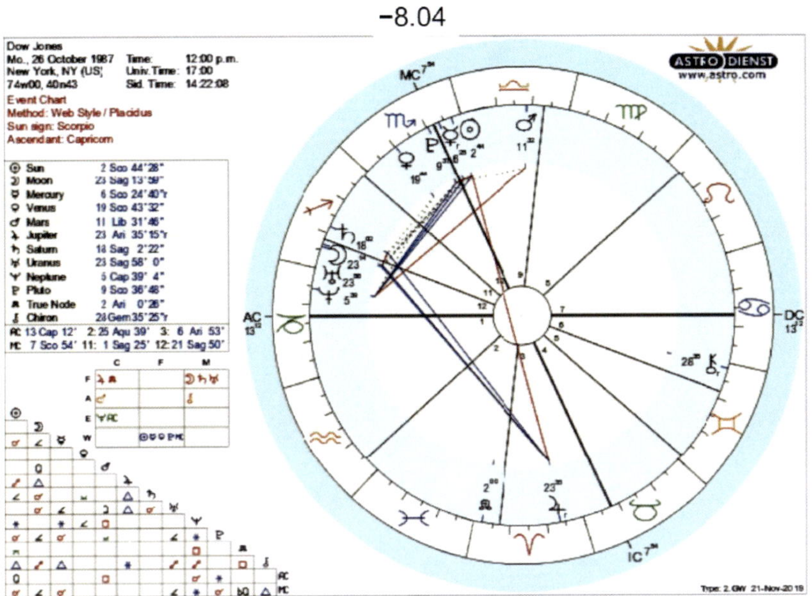

Here is where Mars was positioned in the sky that day

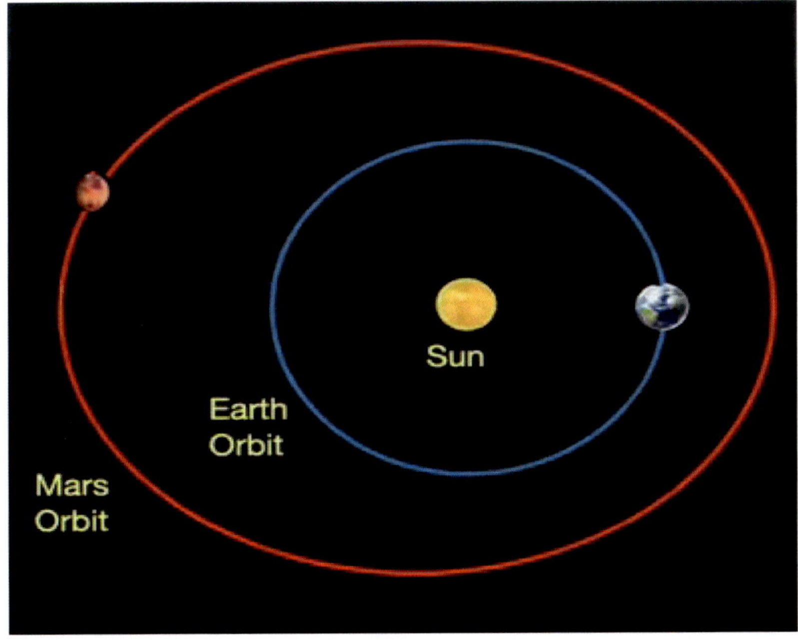

October 15, 2008, the stock market fell -7.87%

−7.87%

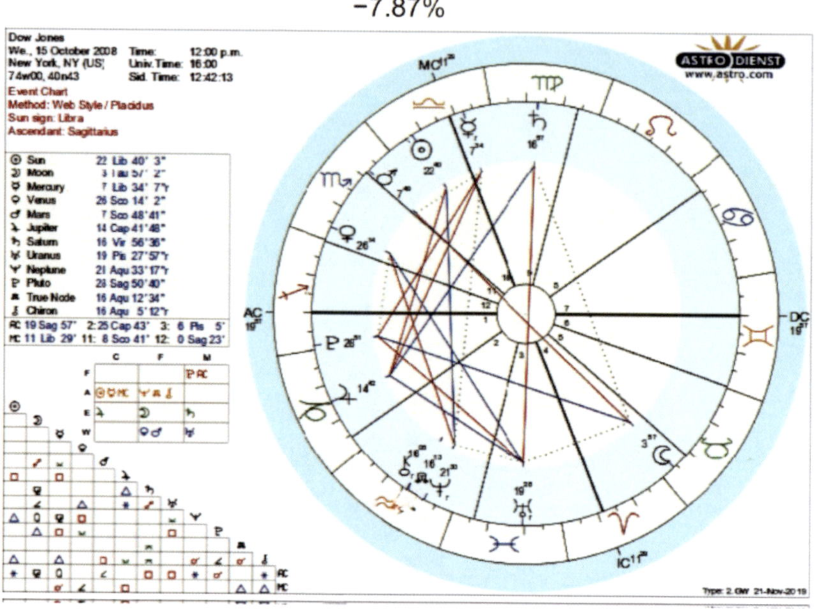

Here is where Mars was positioned in the sky that day

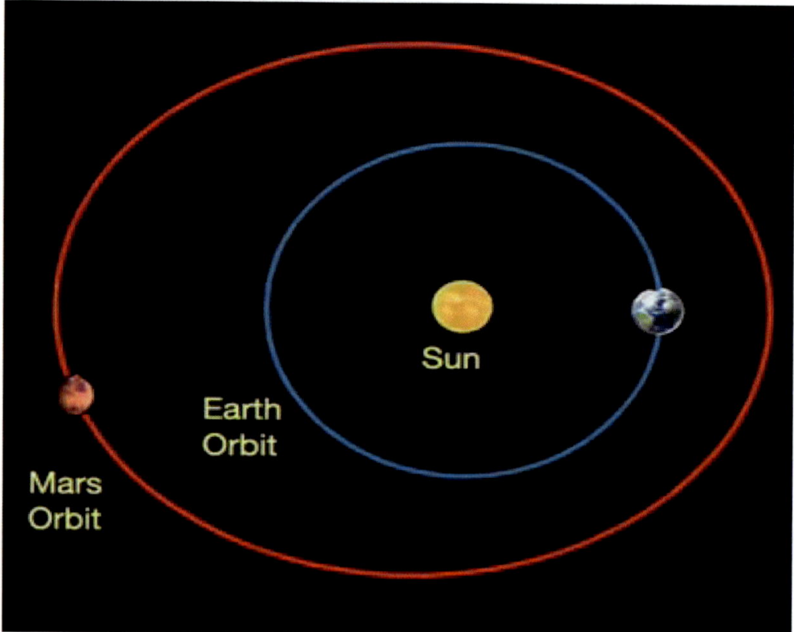

July 21, 1933, The stock market dropped -7.84

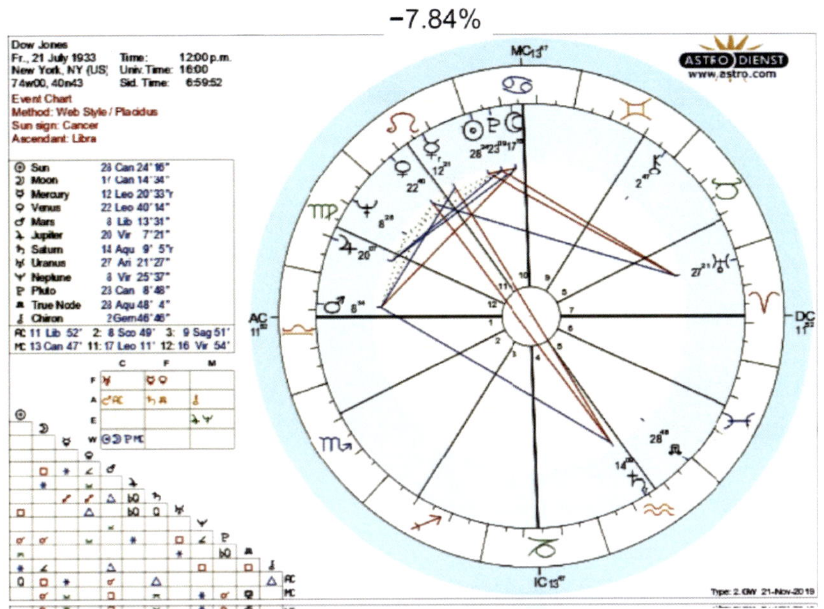

Here is where Mars was positioned in the sky that day

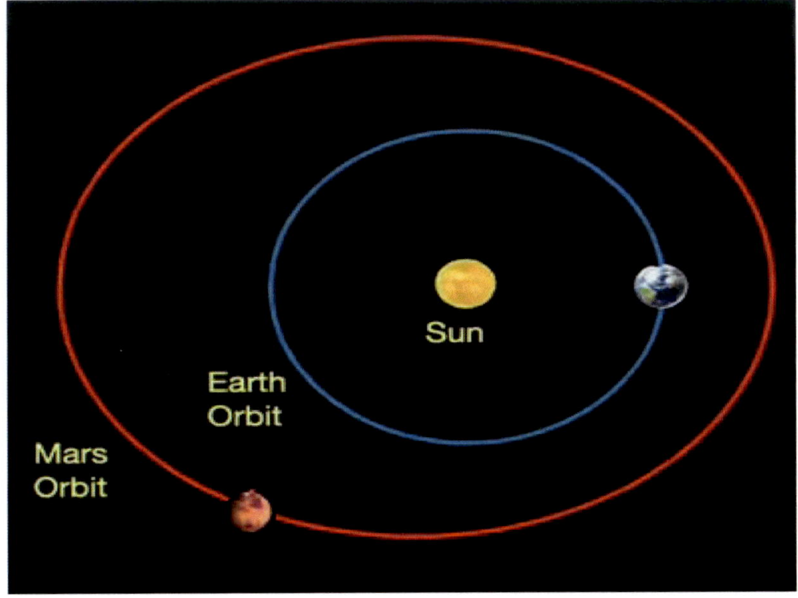

October 18, 1937, the stock market dropped -7.75%

−7.75%

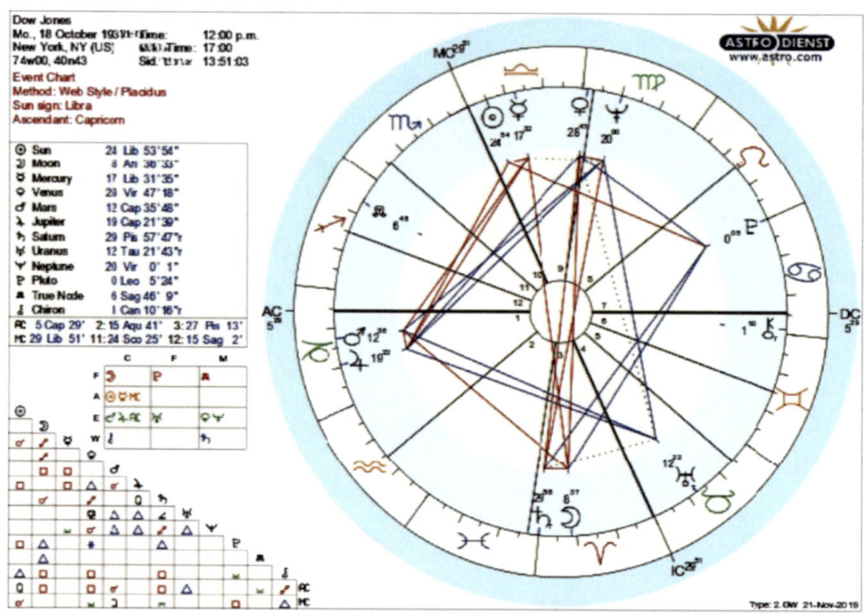

Here is where Mars was positioned on that day

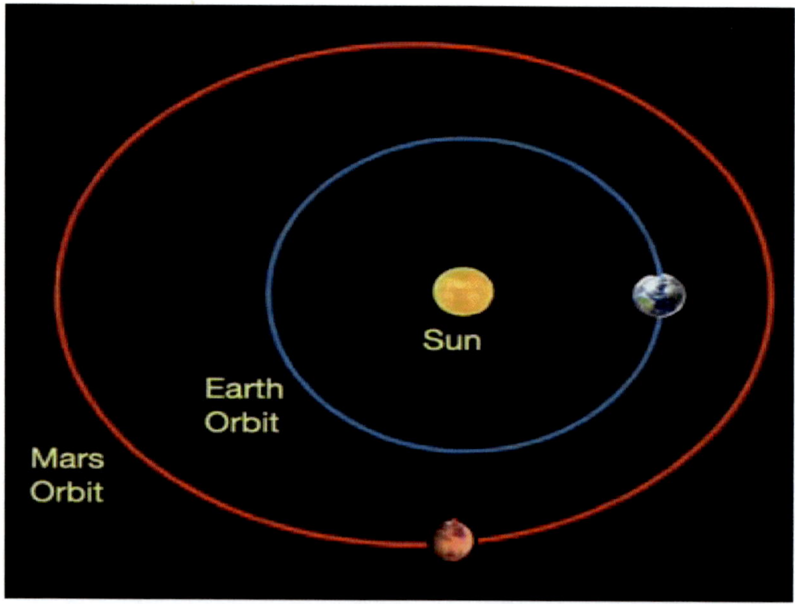

On December 1, 2008, the stock market dropped -7.70%

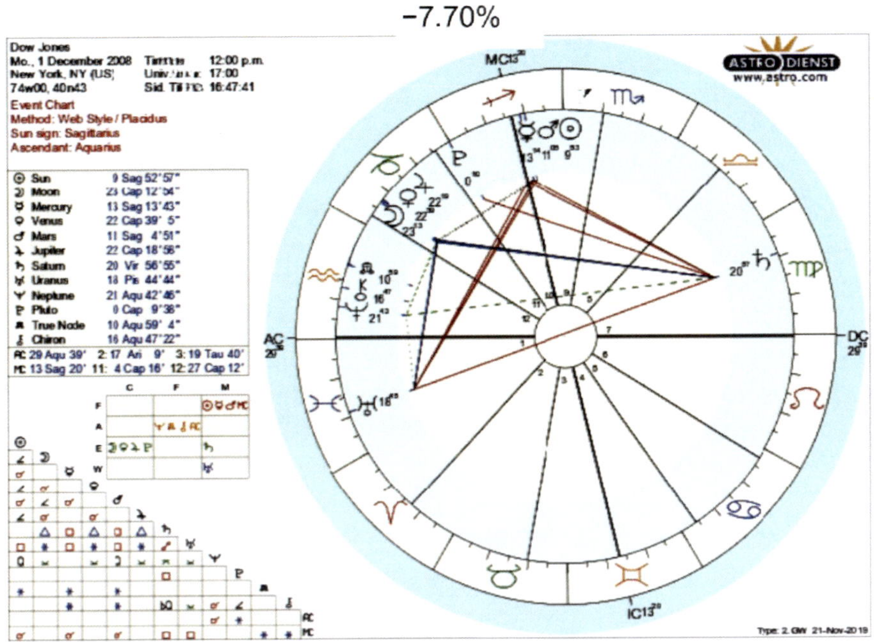

Here is where Mars was positioned in that sky that day

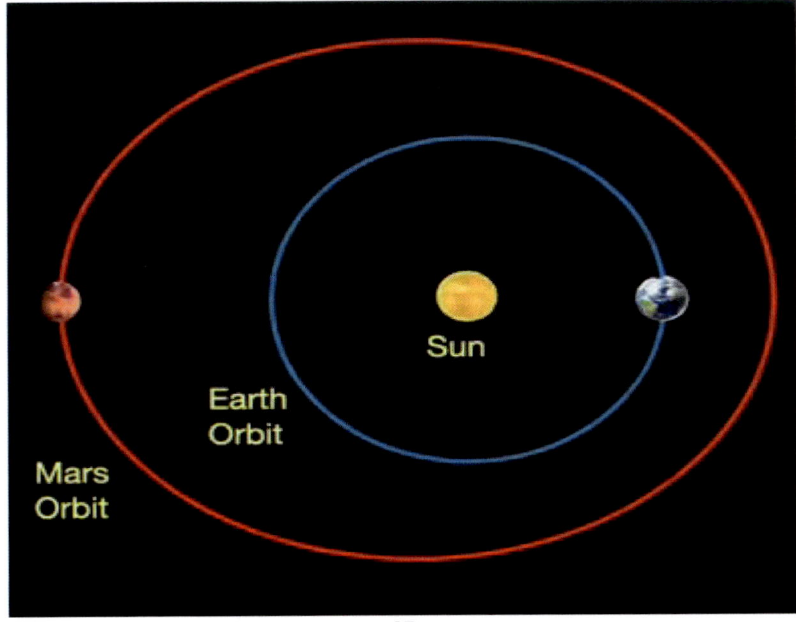

October 9, 2008, the stock market fell -7.33%

−7.33 %

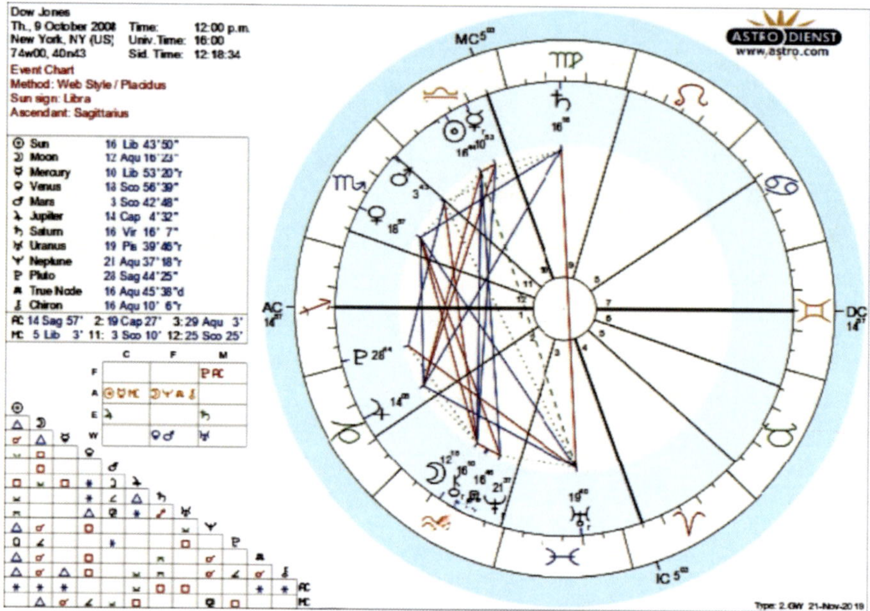

Here is where mars was positioned in the sky on that day

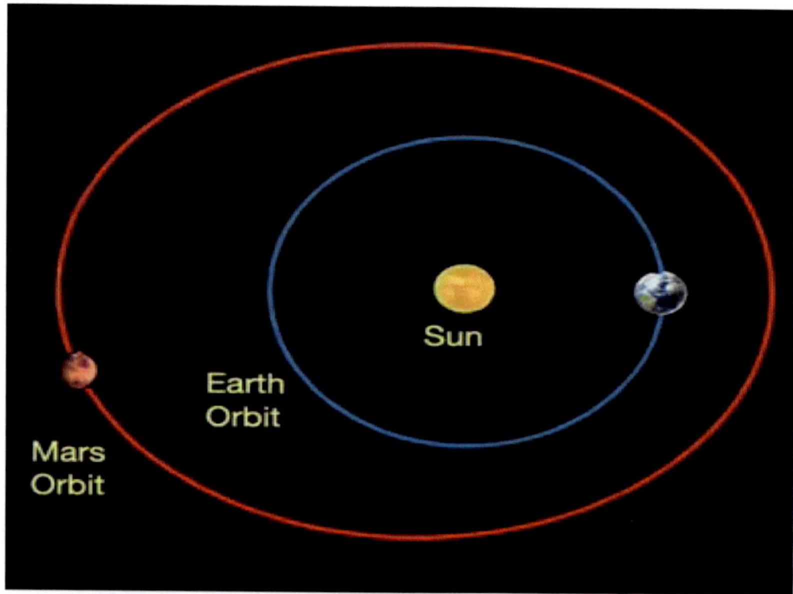

On February 1, 1917, the stock market dropped -7.24%

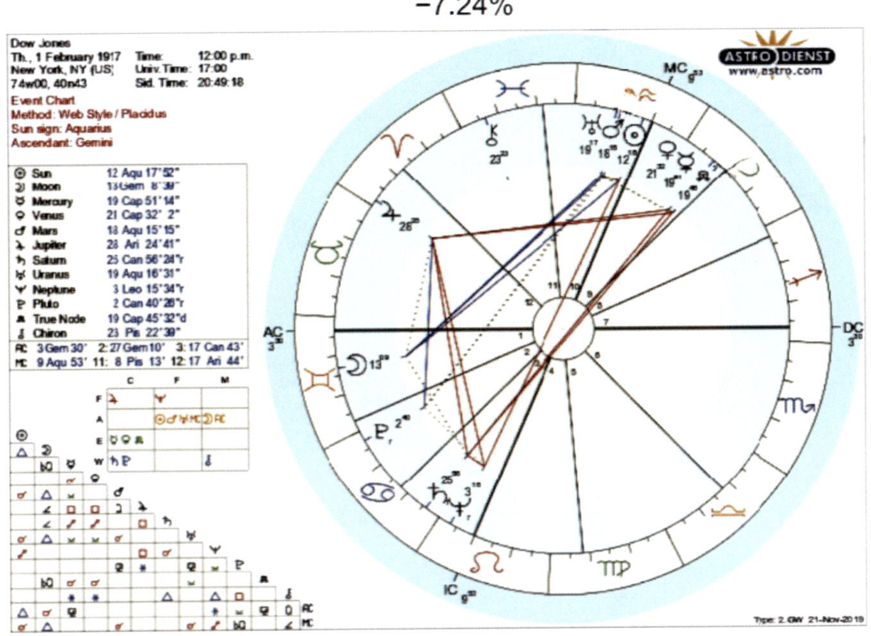

Here is where Mars was positioned in the sky on that day

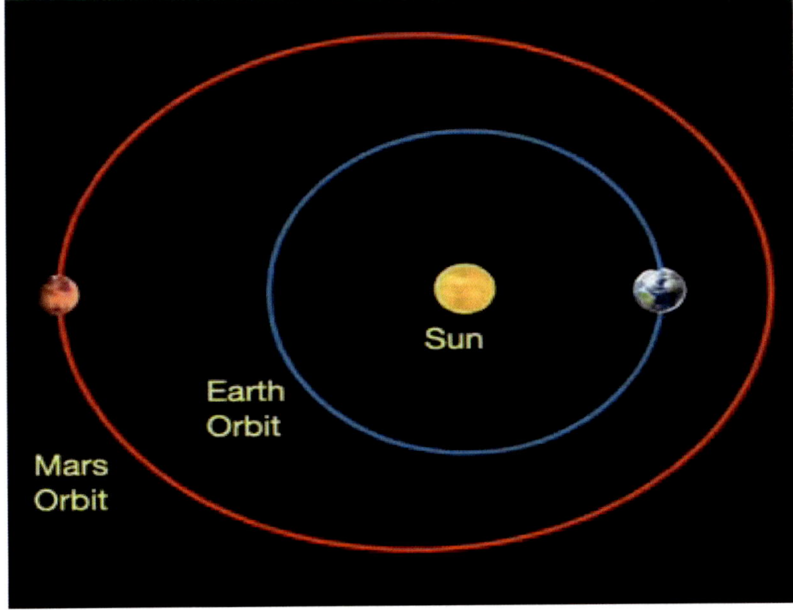

On October 27, 1997, the stock market crashed -7.18%

−7.18%

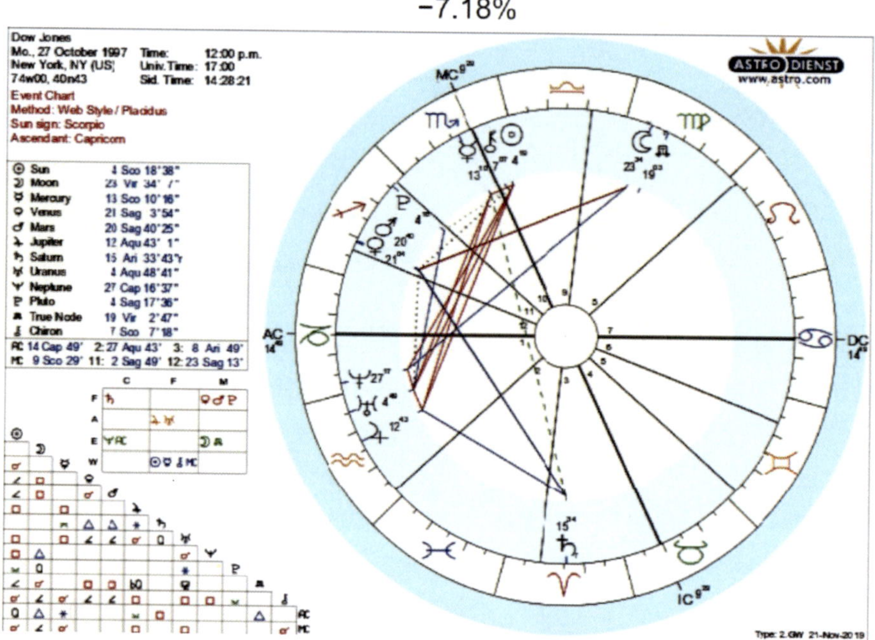

Here is where Mars was positioned in the sky on that day

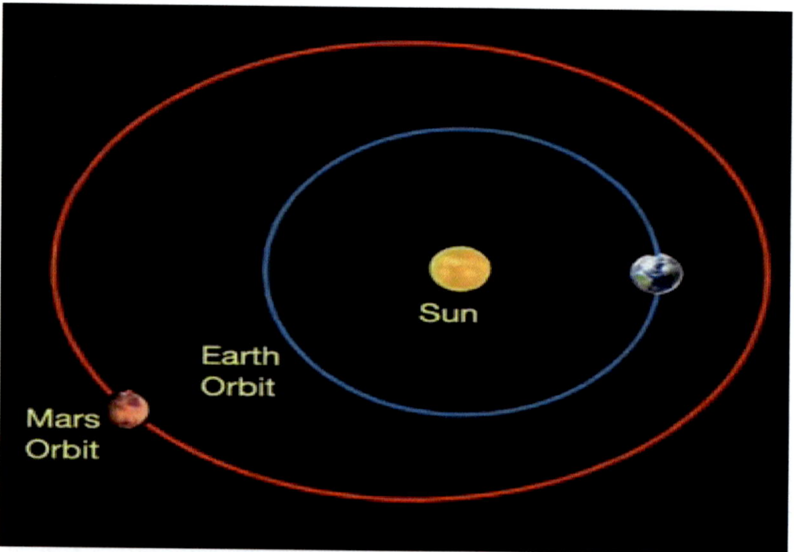

October 5, 1932, the stock market dropped -7.15 percent

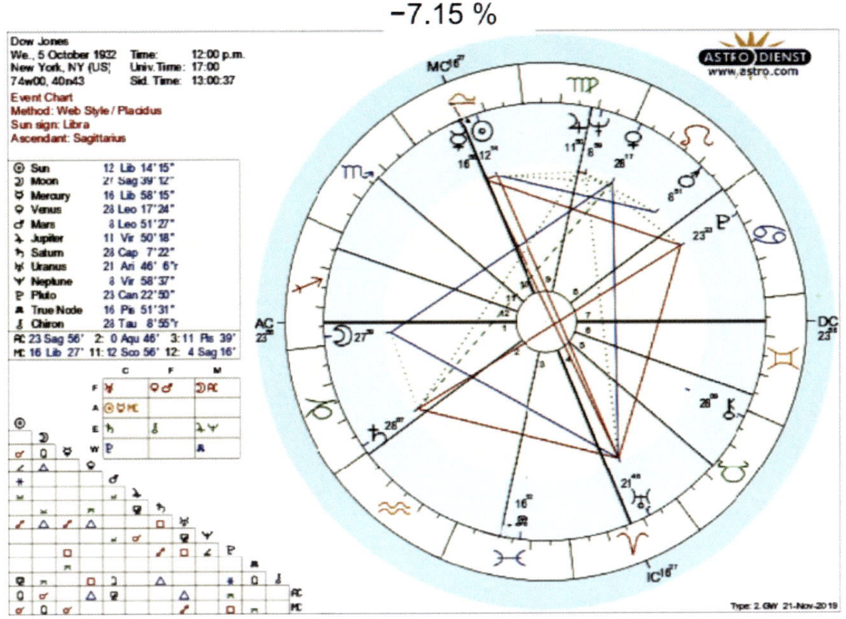

Here is where Mars was positioned in the sky on that day

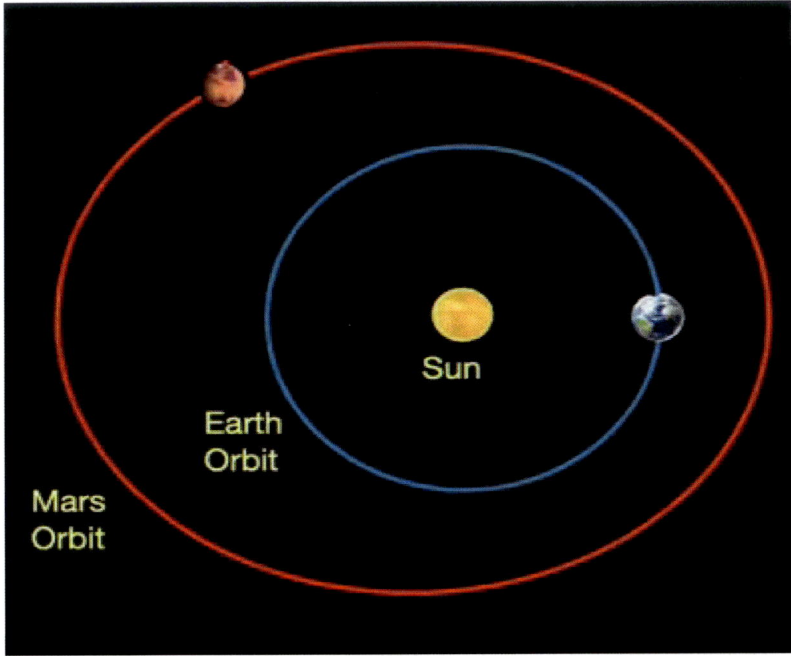

September 17, 2001, the stock market fell -7.13%

−7.13%

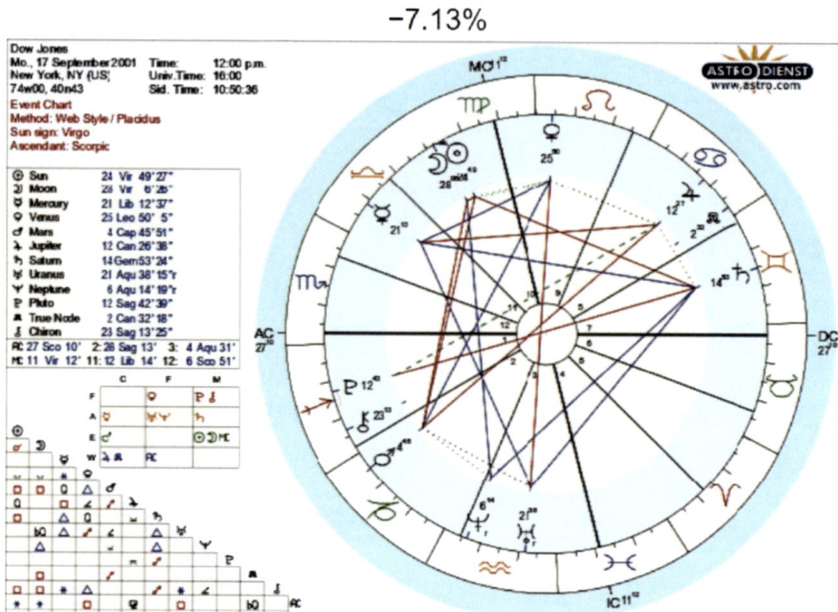

Here is where Mars was positioned in the sky that day

September 24, 1931, the stock market dropped -7.07%

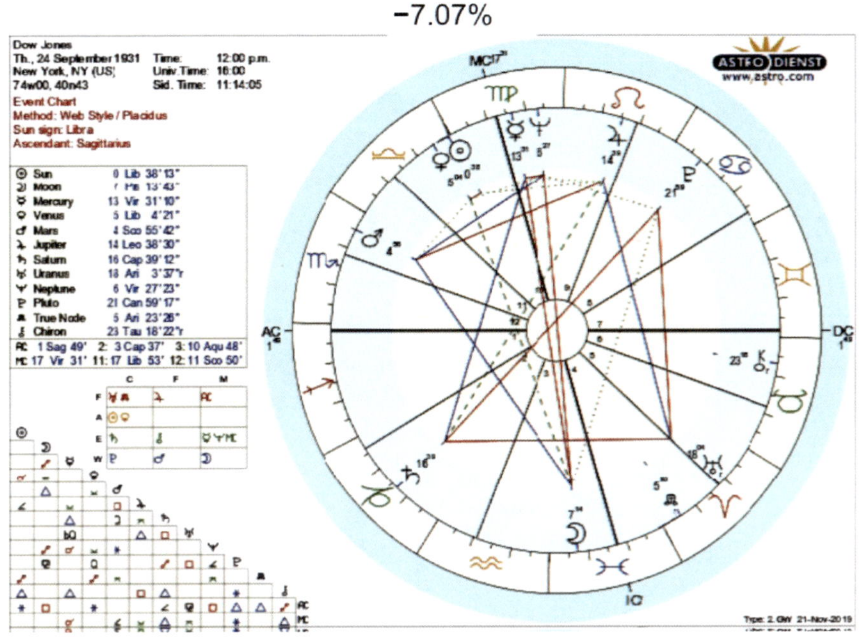

Here is where Mars was positioned in the sky on that day

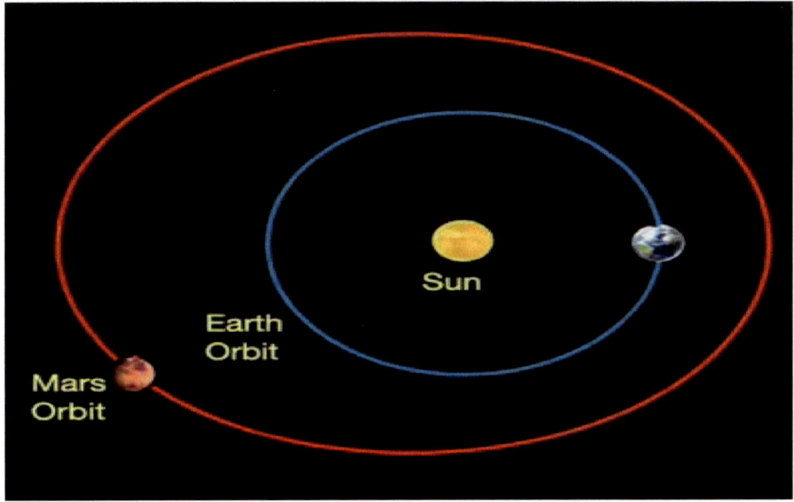

July 20, 1932, the stock market dropped -7.07%

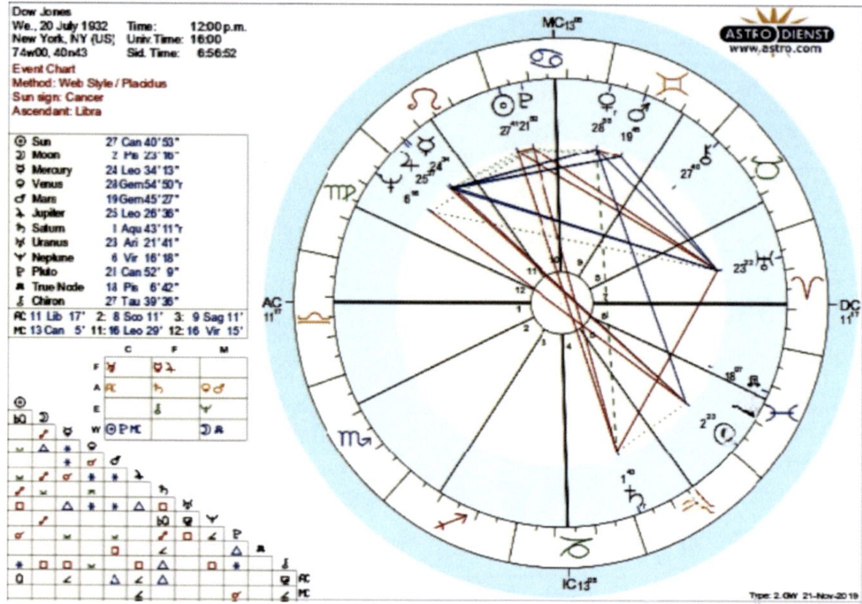

Here is where Mars was positioned in the sky that day

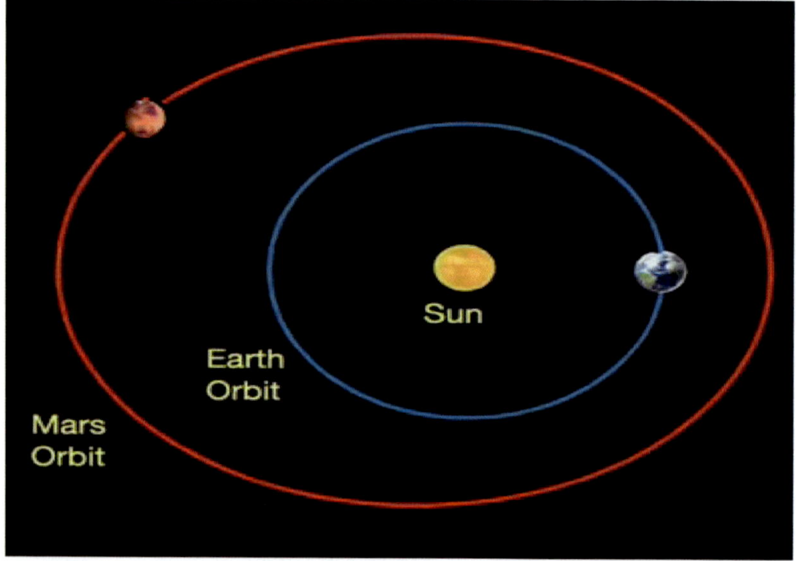

July 30, 1914, the stock market dropped -6.91%

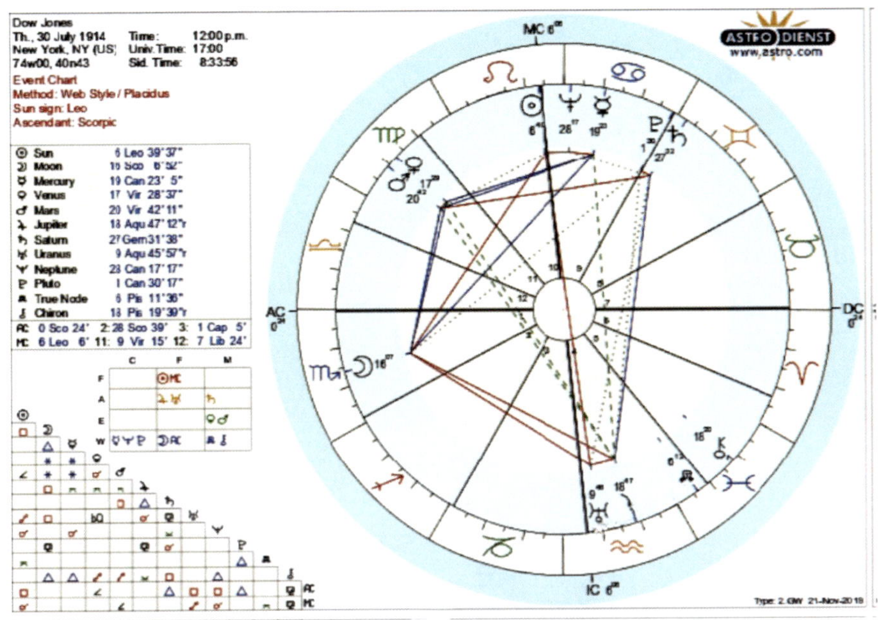

Here is where Mars was positioned in the sky on that day

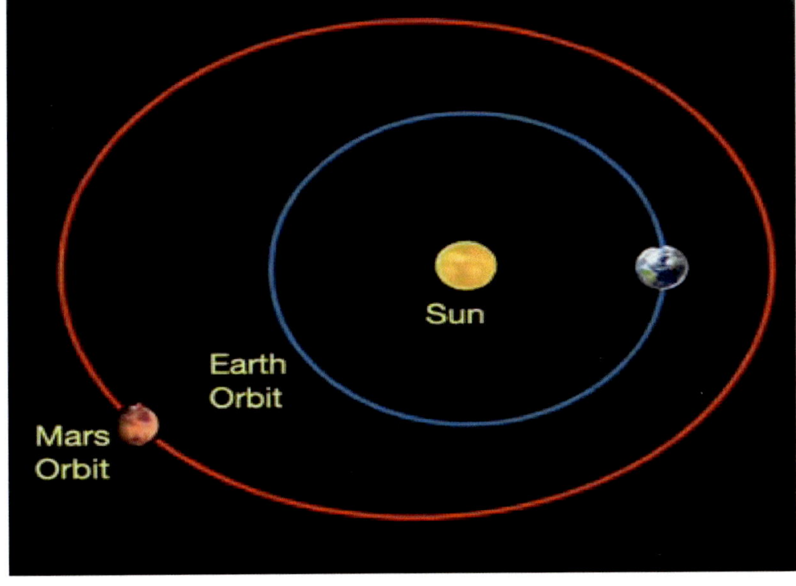

September 29, 2008, the stock market dropped – 6.98%

-6.98%

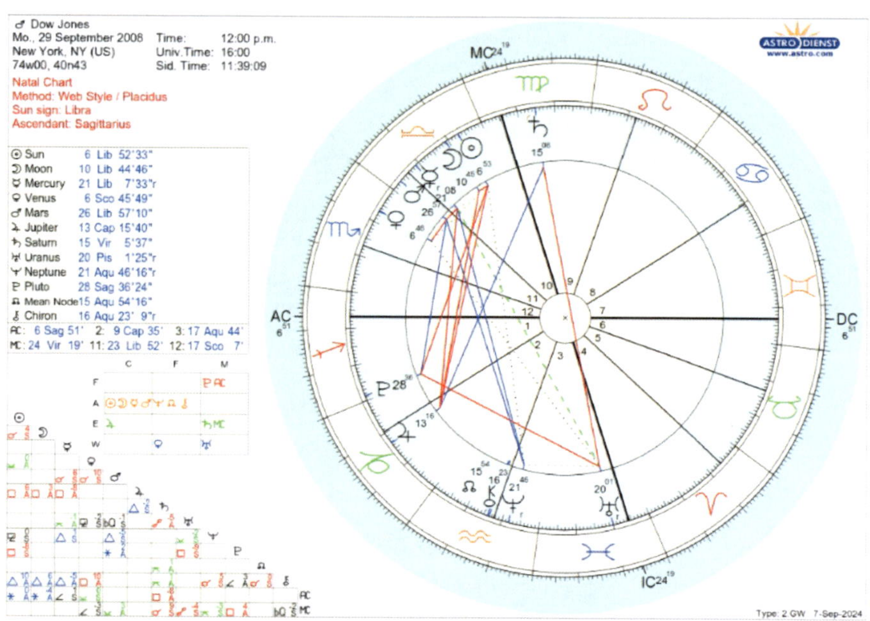

Here is where Mars was positioned in the sky that day

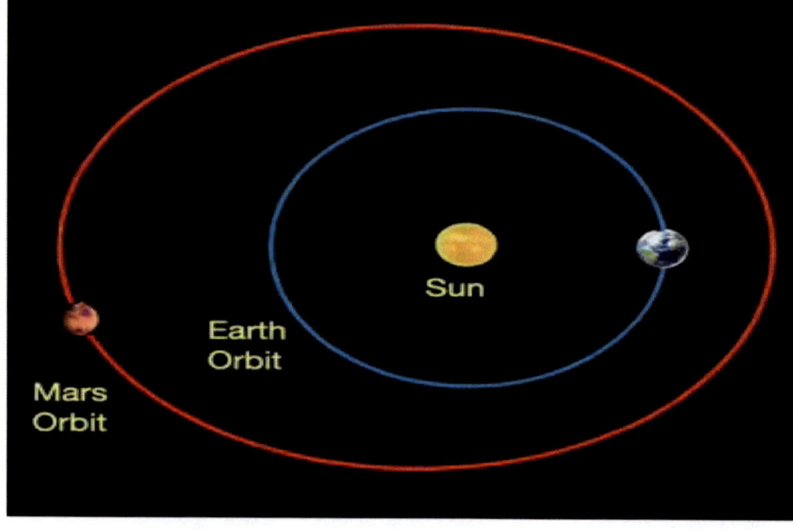

August 8, 2011, the stock market dropped -5.15%

-5.15%

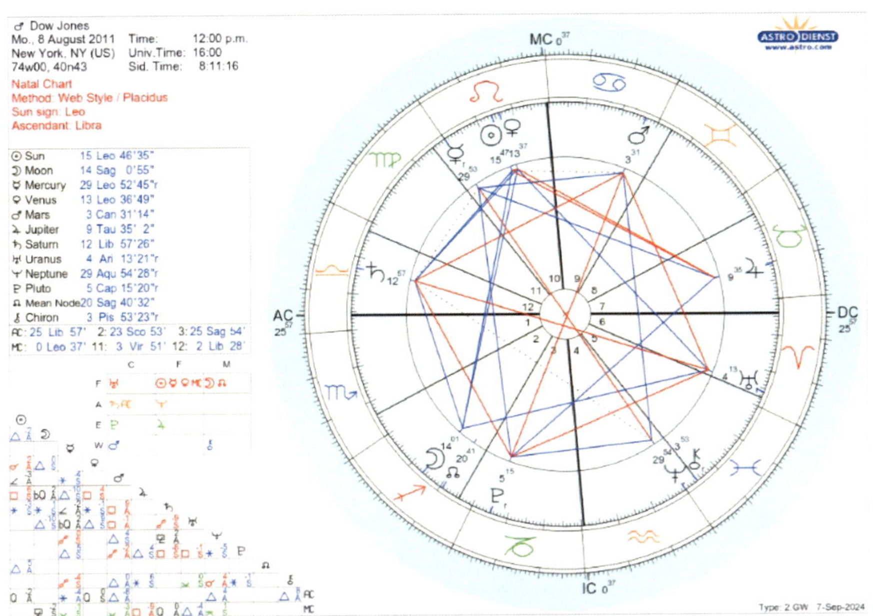

Here is where Mars was positioned in the sky on that day

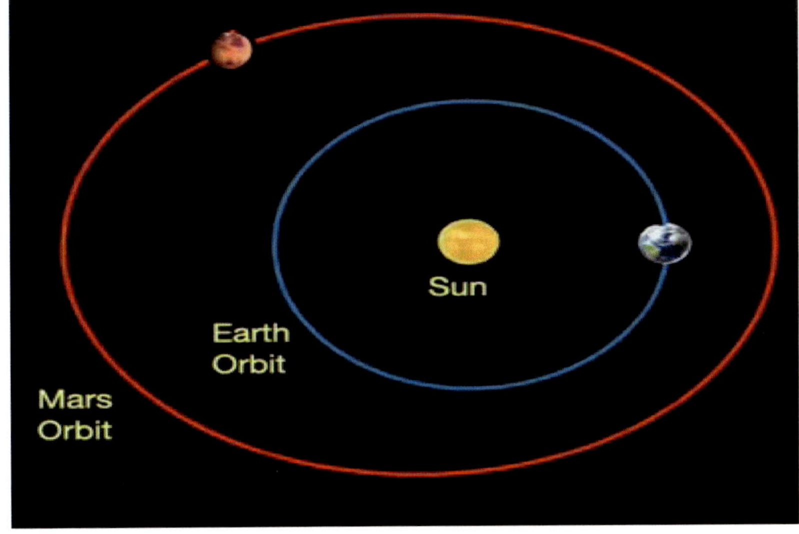

April 14, 2000, the stock market dropped -5.66%

-5.66%

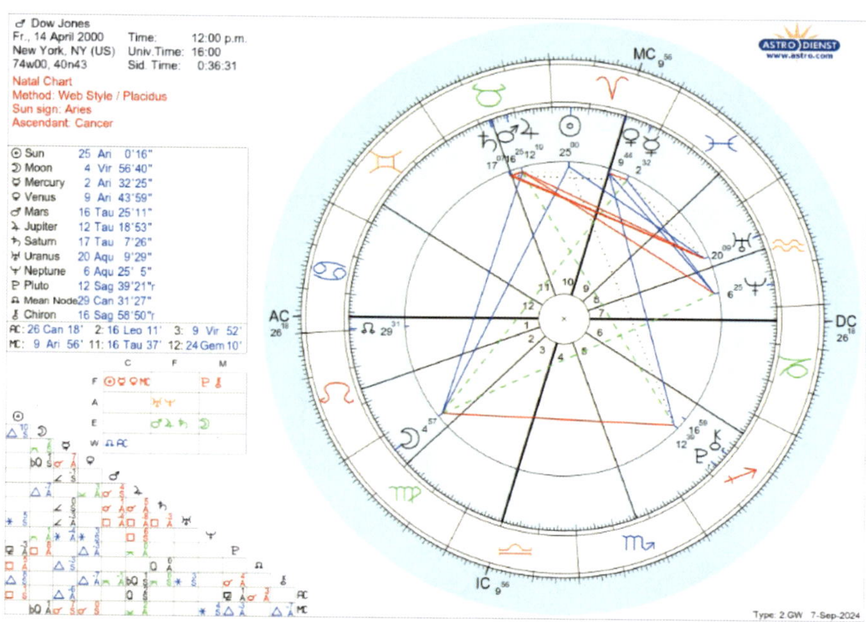

Here is where Mars was positioned in the sky that day

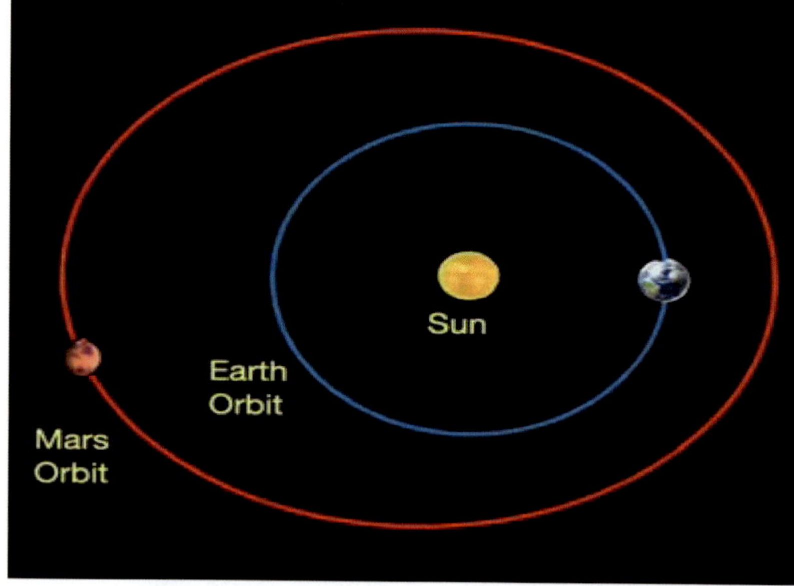

In all the days of major stock market crashes and downturns in the Dow Jones history, Mars was always in the orbital phase, from earth's point of view, marked with the white line.

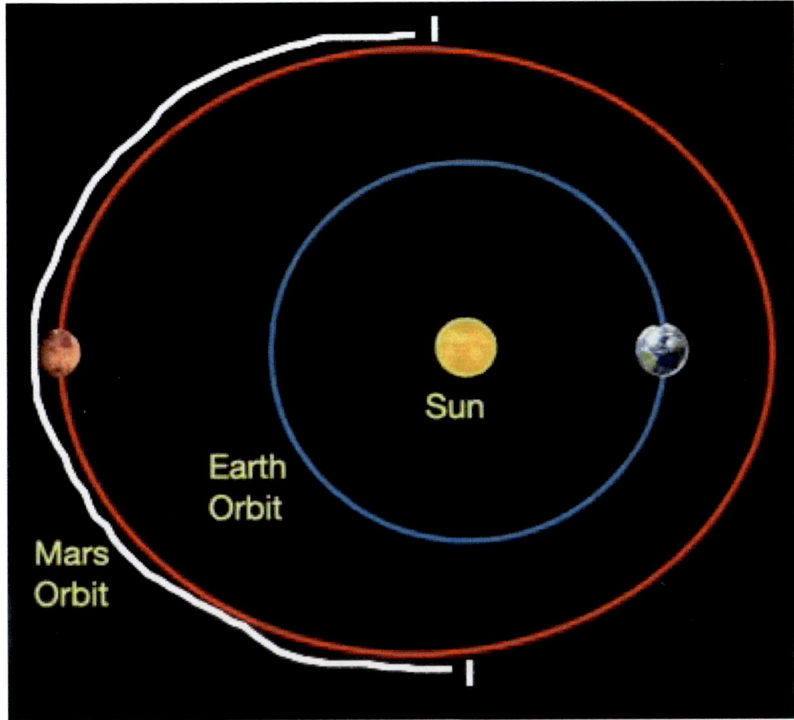

This data shows that a major stock market crash will never happen when Mars is orbiting in the area not marked with the white line. We can say this with 100% certainty.

The white area is the phase of the orbit when Mars is going further out from earth, but also when it's gravity is puling Earth's axial tilt towards the sun, possibly bringing warmer temperatures, which should affect investor sentiment most negatively, presuming that warmer temperatures relative to the mean affect cognitive function and trigger some variant of irritability or pessimism. There are studies that corroborate this dynamic between warmer temperatures and negative mood states.

Outside of the white area, as Mars gets closer to earth, Mars's gravity is puling earth's axial tilt away from the sun, bring presumably cooler temperatures, and less negative mood outcomes, which may explain why major stock market crashes never happen during that phase of Mars's orbit

Dow Jones percentage changes between 1896 and 2023, in correlation with the orbital phase of Mars.
By Anthony of Boston

Please refer to this paper for context before analyzing the data below
https://www.academia.edu/123648970

Below are the time-frames of when Mars was going behind the sun from Earth's point of view, along with how the Dow Jones performed during those times. All 25 major stock market crashes occurred when Mars was anywhere along the white line. This is indicated in the data in parenthesis. The theory is that when Mars is in this orbital phase, its gravity pulls earth's axial tilt toward the sun, increasing warmth and negatively affecting investor sentiment.

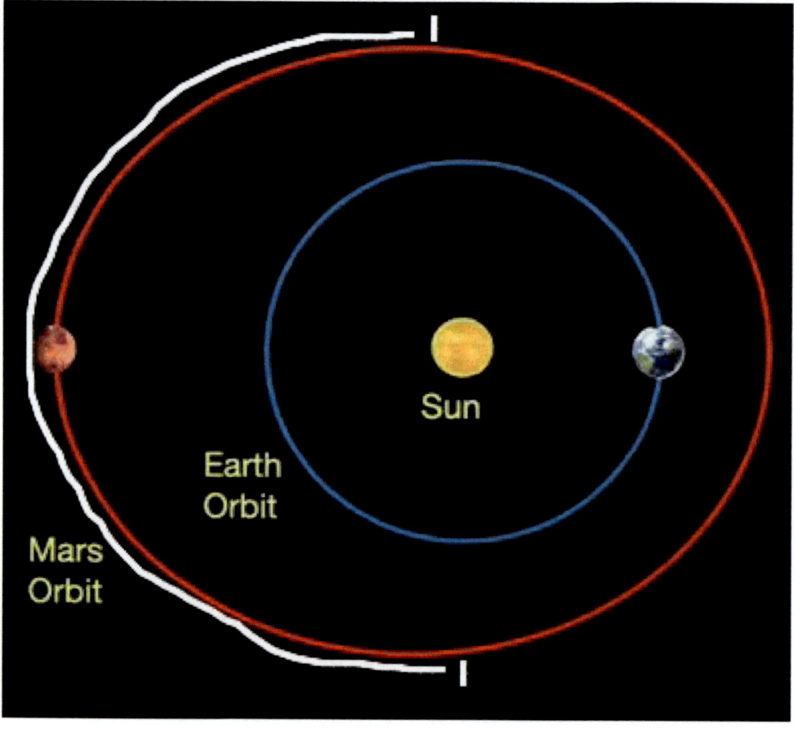

July 15, 1896 to September 1, 1896, Mars was behind the sun. The Dow Jones fell -3.46%
Feb 15, 1897- October 18 1898, Mars was behind the sun. The Dow Jones rose +29%
Mar 24 1899 – Nov 15, 1900, Mars was behind the sun. The Dow Jones fell -3.14%
(Dec 18, 1899, the stock market fell -8.72%)
May 2 1901 - Dec 27 1902, Mars was behind the sun. Dow dropped -14.93%
June 9 1903 - January 26, 1905, Mars was behind the sun, the Dow rose +23%

Aug 1 1905 – March 14, 1907, Mars was behind the sun, the Dow dropped -4.84%
(March 14, 1907, the stock market fell -8.29%)
Oct 14 1907 – May 13 1909 – Mars behind the sun, Dow rose +40%
Dec 18 , 1909 – Aug 8, 1911 – Mars behind the Sun, Dow dropped -15.63%
Jan 27, 1912 – Oct 4, 1913 – Mars behind the Sun,Dow dropped -0.30%
March 11, 1914 – Nov 9, 1915 – Mars behind the sun, Dow rose +17.29%
(July 30, 1914, the stock market dropped -6.91%)
Apr 18, 1916 – Dec 12, 1917 – Mars behind the Sun, Dow dropped -26.10%
 (On February 1, 1917, the stock market fell -7.24%)
May 22, 1918 – Jan 14, 1920 - Mars behind the sun, Dow rose +23.54%
Jul 7, 1920 – Feb 19, 1922 – Mars behind sun, Dow dropped -7.60%
Sept 15, 1922 – April 12, 1924 – Mars behind the sun, Dow dropped -8.78%
Nov 23, 1924 – July 10, 1926 – Mars behind sun, Dow rose +36%
Jan 16, 1927 – Sept 15, 1928 – Mars behind the sun, Dow rose 45%
Mar 1, 1929 – Oct 27, 1930 - Mars behind the sun, Dow dropped -38%
(October 29, 1929 Stock Market crash)
(November 6 1929 Stock Market Crash)
Mar 28, 1931 – Nov 30, 1932- Mars behind the sun, Dow dropped -92.36%
(August 12, 1932, stock market dropped -8.4 %)
(September 24, 1931, the stock market dropped -7.07%)
(October 5, 1932, the stock market dropped -7.15 %)
(July 20, 1932, the stock market dropped -7.07%)
May 3, 1933 – Jan 1, 1935, Mars behind the sun, Dow rose +37%
(July 21, 1933, The stock market dropped -7.84%)
June 16, 1935 – Feb 3, 1937, Mars behind the sun, Dow rose +47.67%
Aug 10, 1937 – Mar 22, 1939, Mars behind the sun, Dow dropped -21.97%
(October 18, 1937, the stock market dropped -7.75%)
Oct 25, 1939 – Jun 2, 1941 – Mars behind the sun, Dow dropped -25.90%
Dec 31, 1942 – Aug 28, 1943 – Mars behind the sun, Dow rose +21.05%
Feb 13, 1944 – Oct 12, 1945 – Mars behind the sun, Dow rose +32%
Mar 17, 1946 – Nov 18, 1947 – Mars behind the sun, Dow dropped – 4.61%
Apr 23, 1948 – Dec 18, 1949 – Mars behind the sun, Dow rose +8.78%
May 29, 1950 – Jan 22, 1952- Mars behind the sun, Dow rose +22.91%
Jul 16, 1952 – Mar 2, 1954 – Mars behind the sun, Dow rose +7.72%
Oct 1, 1954 - April 28, 1956 – Mars behind the sun, Dow rose +36.53%
Dec 7 1956 – July 25 1958 – Mars behind the sun, Dow rose +2.42%
Jan 27, 1959 – Sep 26, 1960 – Mars behind the sun, Dow dropped -1.68%
Mar 5, 1961 – Nov 6, 1962 – Mars behind the sun, Dow dropped -7.97%
Apr 10 1963 – Dec 6, 1964 – Mars behind the sun, Dow rose +21.51%
May 17, 1965 – Jan 8, 1967 – Mars behind the sun, Dow dropped -14.02%
Jun 26, 1967 – Feb 14, 1969 – Mars behind the sun, Dow rose +8.68%
Aug 24, 1969 - Apr 1, 1971 – Mars behind the sun, Dow rose +9.19%
Nov 17, 1971 – Jun 20, 1973 – Mars behind the sun, Dow rose +8.88%
Jan 15, 1974 – Sept 9 1975 – Mars behind the sun, Dow rose +2.03%
Feb 21, 1976 – Oct 20, 1977 – Mars behind the sun, Dow dropped -18.33%
Mar 29, 1978 – Nov 23, 1979 – Mars behind the sun, Dow dropped +8.27%
May 1, 1980 – Dec 24, 1981 – Mars behind the sun, Dow rose +8.21%
Jun 9, 1982 – Jan 30, 1984 – Mars behind the sun, Dow rose +44%
Aug 1, 1984 – Mar 16, 1986 – Mars behind the sun, Dow rose +49%
Oct 15, 1986 – May 19, 1988 – Mars behind the sun, Dow rose +16.03%
(October 19, 1987 Stock market crash)
(October 26, 1987, the stock market dropped -8.04%)
Dec 17, 1988 – Aug 7 1990 – Mars behind the sun, Dow rose +24.76%
Feb 6, 1991 – Oct 5, 1992 – Mars behind the sun, Dow rose +14.44%

Mar 12, 1993 – Nov 10, 1994 – Mars behind the sun, Dow rose +10.79%
Apr 21, 1995 – Dec 13 1996 – Mars behind the sun, Dow rose +40.92%
May 20, 1997 – Jan 14 1999 – Mars behind the sun, Dow rose +26.49%
(On October 27, 1997, the stock market crashed -7.18%)
Jul 8, 1999 – Feb 22, 2001 – Mars behind the sun, Dow dropped -3.20%
(April 14, 2000, the stock market dropped -5.66%)
Sep 13 2001 – Apr 18, 2003 – Mars behind the sun, Dow dropped -9.25%
(September 17, 2001, the stock market fell -7.13%)
Dec 1, 2003 – July 12, 2005 – Mars behind the sun, Dow rose +8.13%
Jan 17, 2006 – Sept 17, 2007 – Mars behind the sun, Dow rose +21.00%
Feb 27, 2008 – Oct 28, 2009 – Mars behind the sun, Dow dropped -16.66%
(October 15, 2008, the stock market fell -7.87%)
(September 29, 2008, the stock market dropped -6.98%)
(October 9, 2008, the stock market fell -7.33%)
(On Dec 1, 2008, the stock market dropped -7.70%)
Apr 5, 2010 – Dec 1, 2011 – Mars behind the sun, Dow rose +12.35%
(August 8, 2011, the stock market dropped -5.15%)
May 8, 2012 – Jan 2 2014 – Mars behind the sun, Dow rose + 24.46%
Jun 15 2014 – Feb 8 2016 – Mars behind the sun, Dow dropped -2.81%
Aug 15, 2016 – Mar 23, 2018 – Mars behind the sun, Dow rose +24%
Nov 4 2018 – Jun 7, 2020 – Mars behind the sun, Dow rose +7.28%
(March 9, 12 and 16 of 2020, Stock Market Crash)
Jan 1 2021 – Aug 27, 2022 -Mars behind the sun, Dow rose +5.48%

Below are the time-frames of when Mars was passing in front the sun from Earth's point of view, along with how the Dow Jones performed during those times. No major stock market crashes occurred when Mars was anywhere along the white line(see below). This is indicated in the data. The theory is that when Mars is in this orbital phase, its gravity pulls earth's axial tilt away from the sun, increasing cooling and positively affecting investor sentiment.

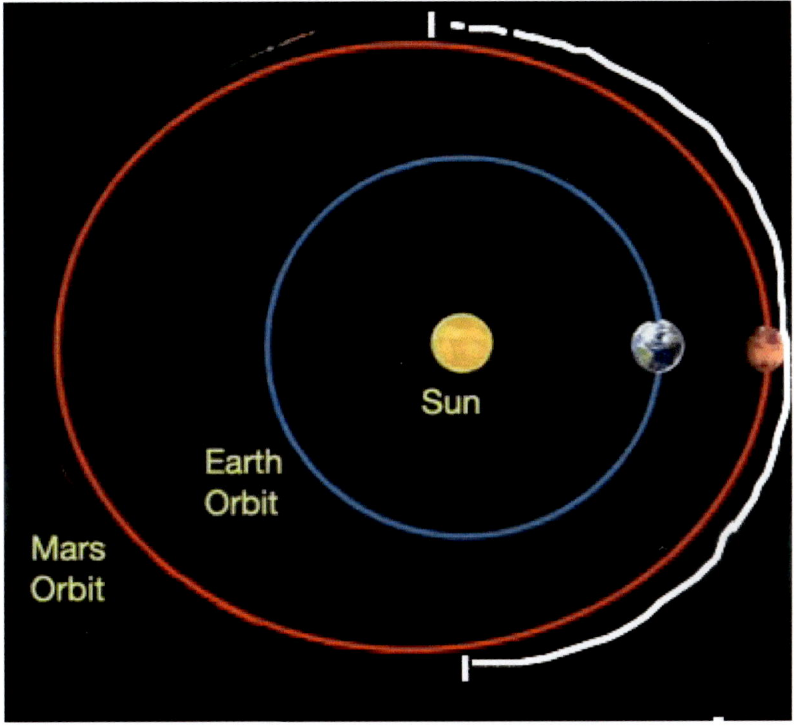

September 2 1896 to Feb 13, 1897, Mars was in front of the sun. The Dow Jones rose +22%
October 19, 1898 - Mar 23 1899, Mars in front of sun, The Dow Jones rose +32.90
November 16, 1900 – May 1, 1901, Mars was in front of the sun. The Dow Jones rose + 13.49%
December 28 1902 - June 9, 1903, Mars was in front of the sun, the Dow dropped -10.09%
January 27, 1905 – July 31, 1905, Mars was in front of the sun, the Dow Jones rose +16%
March 15 1907 – Oct 13 1907 – Mars in front of the sun Dow dropped -18.70%
May 14, 1909 – Dec 17, 1909 - Mars in front of Sun, Dow rose +8.23
Aug 9, 1911 – Jan 26, 1912 – Mars in front of Sun, Dow dropped -0.53%
October 5, 1913 - March 9, 1914 – Mars in front of Sun, Dow rose +1.17%
Nov 10, 1915 – Apr 18, 1916 – Mars in front of Sun Dow rose + 0.04%
Dec 13, 1917 – May 21, 1918 – Mars in front of sun, Dow rose +21%
Jan 15, 1920 – Jul 6, 1920 – Mars in front of sun, Dow dropped -8.06%
Feb 20, 1922 – Sept 14, 1922 – Mars in front of Sun, Dow rose +18.33%
April 13, 1924 – Nov 22,1924 – Mars in front of sun, Dow rose +19.21%

July 11, 1926 – Jan 15, 1927 – Mars in front of sun, Dow rose +0.47%
Sept 16, 1928 – Feb 28, 1929 – Mars in front of sun Dow rose +29%
Oct 28, 1930 – Mar 27, 1931- Mars in front of sun, dow dropped -7.81%
Nov 30, 1932 – May 2, 1933, Mars in front of the sun, Dow rose +35%
Jan 2, 1935 – June 15, 1935, Mars in front of the sun, Dow rose +14.15%
Feb 4, 1937 – Aug 9, 1937, Mars in front of sun, Dow dropped -0.31%
Mar 23, 1939 – Oct 24, 1939 – Mars in front of sun, Dow rose +11.45%
Jun 3, 1941 – Dec 30, 1941 – Mars in front of sun, Dow dropped -3.81%
Aug 29, 1943 – Feb 12, 1944 – Mars in front fo sun, Dow dropped -0.11%
October 13, 1945 – Mar 16, 1946 – Mars in front of the sun, Dow rose +4.87%
Nov 19, 1947 – Apr 22, 1948 – Mars in front of sun, Dow rose +0.91%
Dec 19, 1949 – May 28, 1950 – Mars in front of sun, Dow rose +11.49%
Jan 23, 1952 – Jul 15, 1952 – Mars in front of Sun, Dow rose +0.64%
Mar 3, 1954 – Oct 1, 1954 – Mars in front of the sun, Dow rose +19.25%
April 29, 1956 – Dec 7 1956 – Mars in front of sun, Dow dropped -3.00%
July 26 1958 – Jan 26, 1959 - Mars in front of the sun, Dow rose +16.86%
Sep 27, 1960 – Mar 5, 1961 – Mars in front of the sun, Dow rose +15.37%
Nov 7, 1962 – Apr 9, 1963 – Mars in front of the sun, Dow rose +14.72%
Dec 7, 1964 – May 16 1965 – Mars in front of sun, Dow rose +7.70%
Jan 9, 1967 – Jun 25, 1967 – Mars in front of the sun, Dow rose +8.43%
Feb 15, 1969 – Aug 23, 1969 – Mars in front the sun, Dow dropped -12.54%
Apr 2, 1971 – Nov 16, 1971 – Mars in front the sun, Dow dropped -9.42%
June 21, 1973 – Jan 14, 1974 – Mars in front the sun, Dow dropped -3.98%
Sept 10, 1975 – Feb 20, 1976 – Mars in front the sun, Dow rose +18.21%
Oct 21, 1977 – Mar 28, 1978 – Mars in front of the sun, Dow fell -6.84%
Nov 24 1979 – Apr 30, 1980 – Mars in front the sun, Dow rose +1.17%
Dec 25, 1981 – Jun 8, 1982 – Mars in front the sun, Dow dropped -8.11%
Feb 1, 1984 – Jul 30, 1984 – Mars in front the sun, Dow rose -9.10%
Mar 17 1986 – Oct 14, 1986 – Mars in front the sun, Dow rose +1.18%
May 20 1988 – Dec 16, 1988 – Mars in front the sun, Dow rose +10.03%
Aug 8, 1990 – Feb 5, 1991 – Mars in front the sun, Dow rose +3.77%
Oct 6 1992 – Mar 11, 1993 – Mars in front the sun, Dow rose + 8.58%
Nov 11, 1994 – Apr 20, 1995 – Mars in front the sun, Dow rose +10.37%
Dec 14, 1996 – May 19, 1997 – Mars in front the sun, Dow rose +14.24%
Jan 15 1999 – Jul 7, 1999 – Mars in front the sun, Dow rose +21.08%
Feb 23, 2001 – Sep 12, 2001 – Mars in front of sun, Dow dropped -8%
Apr 19 2003 – Nov 30 2003 – Mars in front of the sun, Dow rose +16%
July 13, 2005 – Jan 16, 2006 – Mars in front the sun, Dow rose +4.30%
Sept 18, 2007 – Feb 26, 2008 – Mars in front the sun, Dow dropped -4.71
Oct 29, 2009 – Apr 4 2010 – Mars in front the sun, Dow rose +12.09%
Dec 2, 2011 – May 7 2012 – Mars in front the sun, Dow rose +8.17%
Jan 3 2014 – Jun 14, 2014 – Mars in front the sun, Dow rose +2.27%
Feb 9, 2016 – Aug 14, 2016 – Mars in front of the sun, Dow rose +15.08%
Mar 24, 2018 – Nov 3, 2018 – Mars in front of the sun, Dow rose +7.69%
Jun 8, 2020 – Dec 31, 2020 – Mars in front the sun, Dow rose +16.45%
Aug 28 2022 – Feb 19 2023 – Mars in front the sun, Dow rose +4.78%

To gain relevant context in regards to what this paper is demonstrating, it is important to take into account a recent study published in Nature Communications in March of 2024, roughly 5 years after this idea was first introduced to the public. In that study published in March of 2024, researchers discovered that Mars is exerting a gravitation pull on earth's tilt, exposing earth to warmer temperatures and more sunlight, all within a 2.4 million year cycle. I assert that this allows us to surmise that, even within smaller timeframes, Mars is still exerting a gravitational pull on earth's axial tilt, enough to raise temperatures when Mars travels behind the sun or lower temperatures when it travels in front of the sun, from earth's point of view. This would affect rainfall if other dynamics trigger the temperature perturbations conducive to precipitation

New evidence for an unexpected player in Earth's multimillion-year climate cycles: the planet Mars

Published: March 12, 2024 3:14pm EDT

Dietmar Müller

Authors

Adriana Dutkiewicz
ARC Future Fellow, University of Sydney

Dietmar Müller
Professor of Geophysics, University of Sydney

Slah Boulila
Associate lecturer, Sorbonne Université

Our existence is governed by natural cycles, from the daily rhythms of sleeping and eating, to longer patterns such as the turn of the seasons and the quadrennial round of leap years.

After looking at seabed sediment stretching back 65 million years, we have found a previously undetected cycle to add to the list: an ebb and flow in deep sea currents, tied to a 2.4-million-year swell of global warming and cooling driven by a gravitational tug of war between Earth and Mars. Our research is published in Nature Communications.

Back in 2014, two scientists from the University of Washington studied climate data spanning 15 years and discovered that lunar tallies affect rainfall. Tsubasa Kohyama and his professor John Wallace studied rainfall data spanning 15 years between 1998 and 2012 and found that the position of the moon when it is overhead from our vantage point standing on earth or under foot, air pressure increases, which leads to elevated temperatures, more absorbed moisture and less rainfall. However, the effect was only 1% of all rainfall variations but the data was significant enough to link the position of the moon with rainfall. At rising or setting from our vantage point, rainfall should theoretically be higher. But at the meridian, according to the study, the moon decreases rainfall. The science behind this study is that the moon's gravity pull's earth's atmosphere higher, increasing air pressure. When this happens, the air beneath becomes warmer and able to absorb more moisture. This study allows us to use the position of the moon as our rainfall trigger.

In addition, with the moon understood as having a stabilizing effect on earth's wobble, we can point to the position of the moon relative to Mars as being a momentary adversarial influence against Mars's gravitational pull upon earth's axial tilt, in that when the moon lines up opposite to Mars, it can momentarily shift temperatures away from the current trend that is fostered by Mars's gravitational pull on the earth.

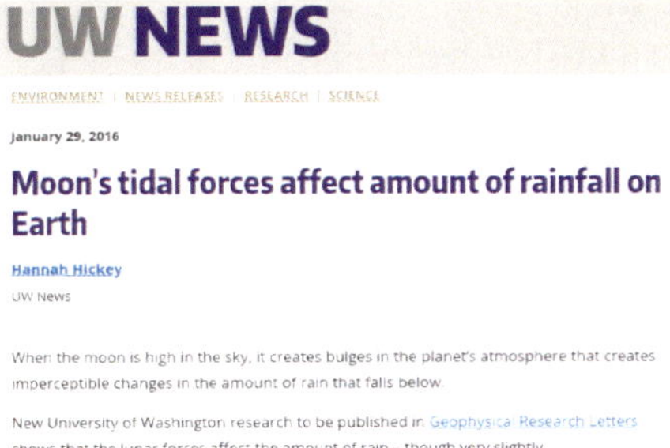

Moon's tidal forces affect amount of rainfall on Earth

ENVIRONMENT | NEWS RELEASES | RESEARCH | SCIENCE

January 29, 2016

Hannah Hickey
UW News

When the moon is high in the sky, it creates bulges in the planet's atmosphere that creates imperceptible changes in the amount of rain that falls below.

New University of Washington research to be published in Geophysical Research Letters shows that the lunar forces affect the amount of rain – though very slightly.

by Anthony of Boston

With this new understanding about the revolution of Mars around the sun and its link to earthly climate patterns and human behavior, we can surmise how this dynamic would play out when it comes to predicting precipitation. The basic premise of precipitation is that warmer air is able to hold onto moisture/water vapor until cooler air comes in and causes the water vapor to undergo a process called condensation, which turns water vapors into liquid droplets or what we know as rain. Understanding how Mars can create the conditions for rain will help us to predict precipitation events much more efficiently. So far, it has been posited that when Mars travels behind the sun, from earth's point of view, its gravitational pull on the earth's axial tilt can expose the earth to more sunlight and warmer temperatures. When Mars travels in front of the sun, from earth's point of view, it's gravitational pull on earth's axial tilt, pulls earth away from the sun, which should trigger less sunlight, less heat and more cooling. With these aspects in mind, we can apply this dynamic to the seasons in which this happens, which would thus allows to predict when warmer air will mix with cooler air or vice versa and create the conditions for moisture to precipitate out and become rainfall.

Here is an example of what I mean. The warmer months in a calendar year are spring and summer which starts around March 20 and lasts until September 20. We can maintain as a constant that this time of the year will have more moisture in the air and less rain, unless afflicted by the Mars variable, such that if Mars is traveling behind the sun during this period, increasing the earth's exposure to sunlight and warmth, it could be expected that rainfall will be less, allowing us to predict that spring and summer that year will be drier. If it's the other way around, that Mars is traveling in the front of the sun during the spring and summer, pulling earth's tilt away, exposing earth to less sunlight and more cooling, we can surmise that the spring and summer will have more precipitation since cooler air brought on by this Mars's configuration will mix with the warmer spring and summer air and create the conditions for precipitation.

This dynamic would also apply when it comes to cooler months, fall and winter between September 20 and March 20th. If Mars is traveling behind the sun during the winter, the warmer air from this will mix with the cooler and create the conditions for precipitation. If Mars is traveling in front of the sun during this period, more cooler air will result with less chance for precipitation.

We can also factor in Mars within 30 degrees of the lunar node as a factor that can exacerbate the conditions for rainfall by pulling and stretching the moon's orbital plane, bringing the moon further away from earth, which has a destabilizing effect on earth's wobble.

With this theoretical framework, we can apply the conditions necessary to trigger actual rainfall. Presuming a period of higher rainfall or lower rainfall based on Mars position relative to earth and the season at the time doesn't provide an actual mechanism that could trigger rainfall. We thus have to envision an scenario where cooler and warmer air will mix in a given period. Lets say that we have Mars traveling behind the sun, during the winter, creating the a scenario for a warmer winter as Mars's gravity pulls earth's axial tilt during this period. In this regard, we can posit that there would be more rain rather than snow during this period. However, we still need to interpolate a scenario where warmer air will mix with cooler air. If this scenario of Mars traveling behind the sun during winter predicts a warmer winter, then in order for there to be rain during that season, a mechanism that brings in cooler air would need to be explained. We can thus insert the lunar scheme.

Back in 2014, two scientists from the University of Washington studied climate data spanning 15 years and discovered that lunar tallies affect rainfall. Tsubasa Kohyama and his professor John Wallace studied rainfall data spanning 15 years between 1998 and 2012 and found that the position of the moon when it is overhead from our vantage point standing on earth or under foot, air pressure increases, which leads to elevated temperatures, more absorbed moisture and less rainfall. However, the effect was only 1% of all rainfall variations but the data was significant enough to link the position of the moon with rainfall. At rising or setting from our vantage point, rainfall should theoretically be higher. But at the meridian, according to the study, the moon decreases rainfall. The science behind this study is that the moon's gravity pull's earth's atmosphere higher, increasing air pressure. When this happens, the air beneath becomes warmer and able to absorb more moisture. This study allows us to use the position of the moon as our rainfall trigger.

In addition, with the moon understood as having a stabilizing effect on earth's wobble, we can point to the position of the moon relative to Mars as being a momentary adversarial influence against Mars's gravitational pull upon earth's axial tilt, in that when the moon lines up opposite to Mars, it can momentarily shift temperatures away from the current trend that is fostered by Mars's gravitational pull on the earth. If we are in a warmer than usual season because Mars is traveling behind the sun, pulling earth's axial tilt toward the sun, we can posit that when the moon lines up opposite to Mars, but behind the earth, we can predict that the moon's gravity pulling the earth's tilt away from the sun will cause a momentary shift in temperatures, which would create the conditions for cooler air to mix in with warmer air and break apart water vapors, allowing water to precipitate out and become rain.

Here is a general idea on how to envision this scenario causing rain.

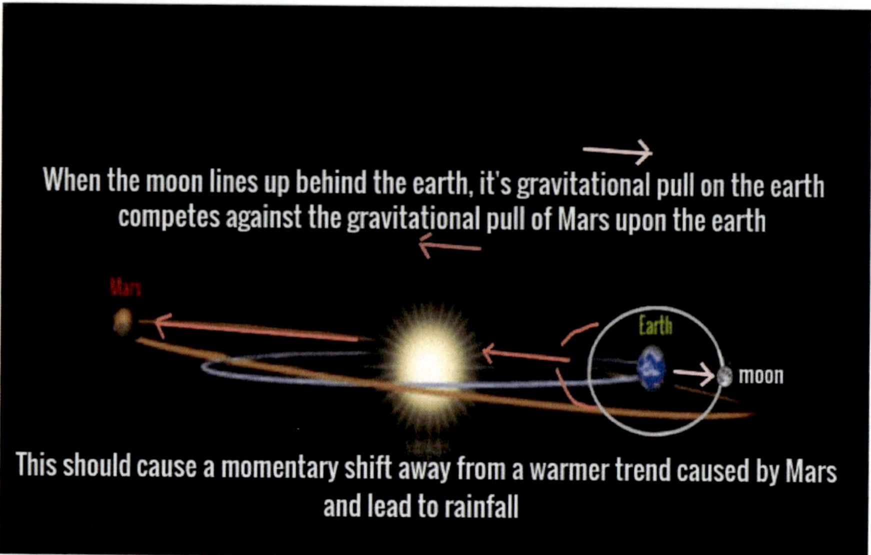

In the image, we see the conditions that could lead to moisture and water vapor absorbed during the warmer trend, later becoming precipitation as the moon interrupts the warmer trend by trying to oppose Mars's gravitational pull and bring earth's tilt away from the sun. This would be momentary, over the course of 1-5 days since the moon travels much faster around the earth than Mars does around the sun.

Now keep in mind there are many variations of this dynamic that could trigger rainfall. For instance, Mars traveling in front of the sun during a summer, thereby causing lower than average temps as Mars's gravity pulls earth's axial tilt away from the sun, could be met with opposition when the moon travels in front of the earth, which would create conditions for precipitation since the moon's gravitational influence on the earth, pulling the earth toward the sun, could interrupt a cooler trend. Warm air would merge with the cooler air, leading to the break up of water vapors. Here is an example that represents such a scenario.

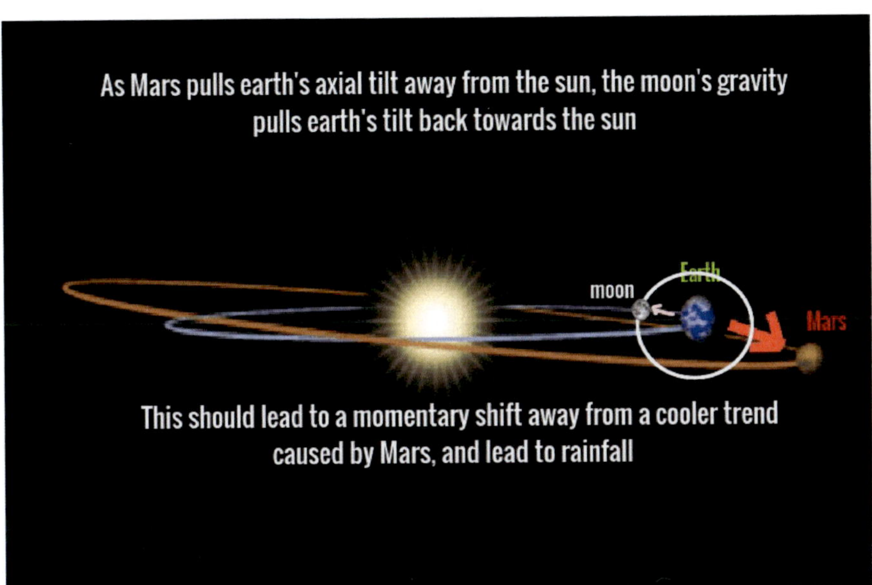

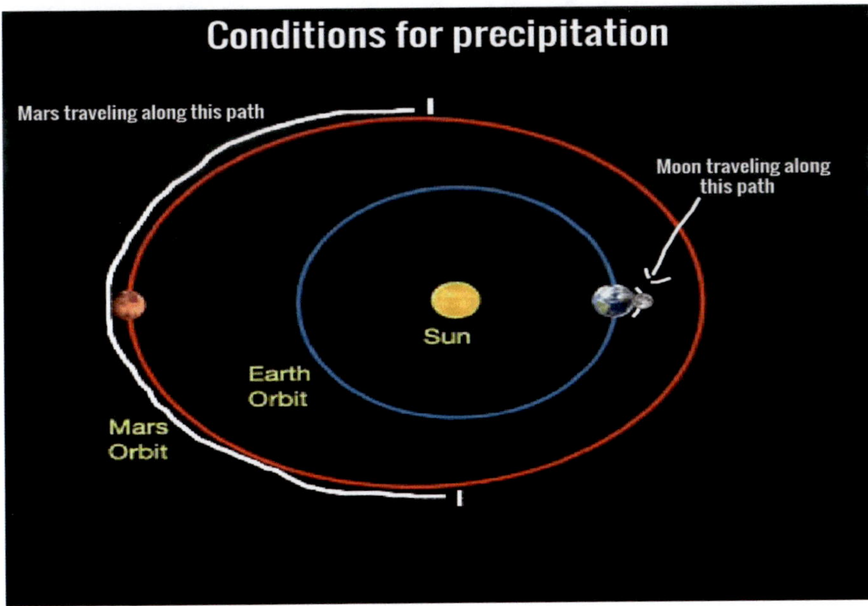

Here is a basic overview for the conditions for precipitation

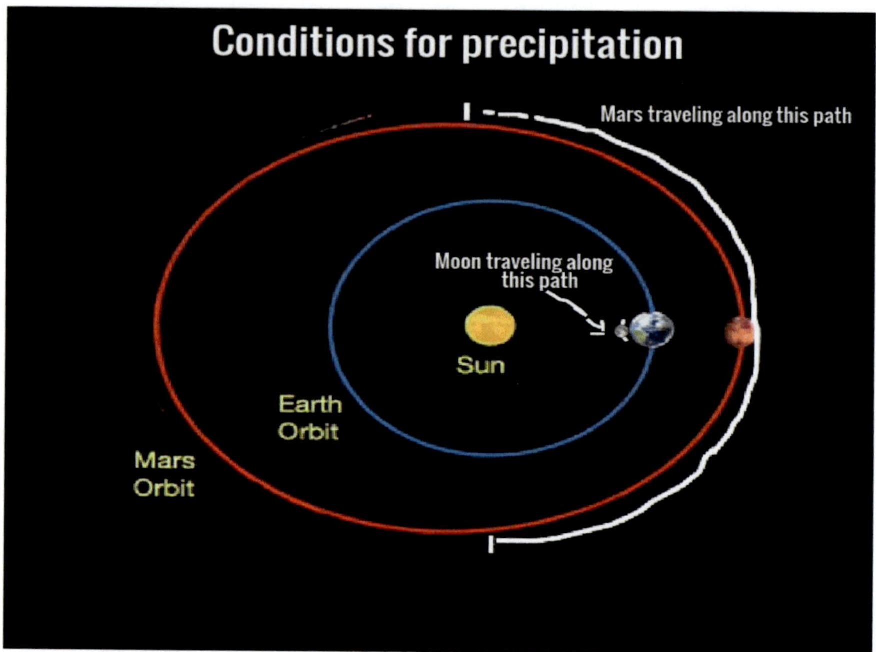

These first two examples interpolate how rain can be fostered by this alignment, and narrows down the parameters that could trigger rainfall. We can now narrow things down further and insert the notion that the closer the alignment at opposition between the moon and Mars, the more likely heavy rain will come about. So now lets narrow down the required path of the moon and Mars.

Conditions for precipitation

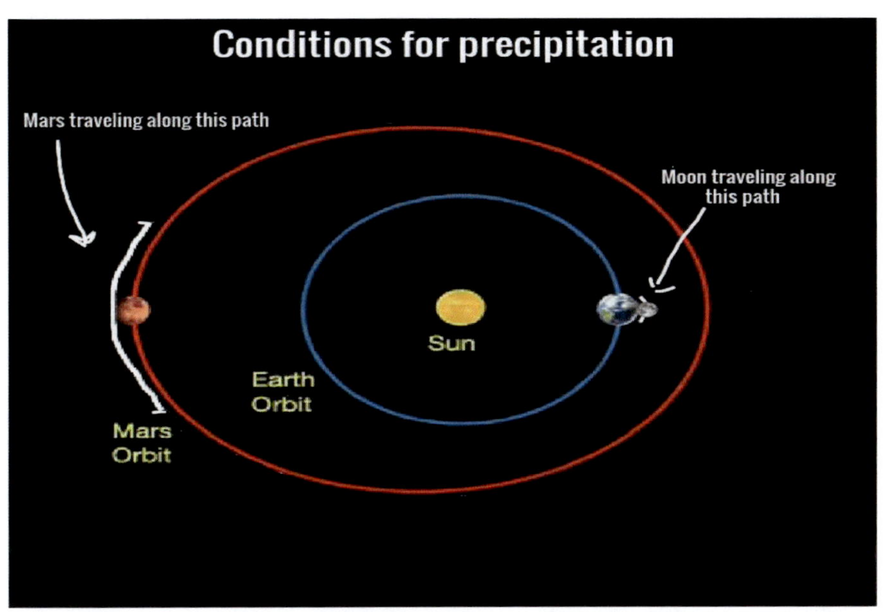

Conditions for precipitation

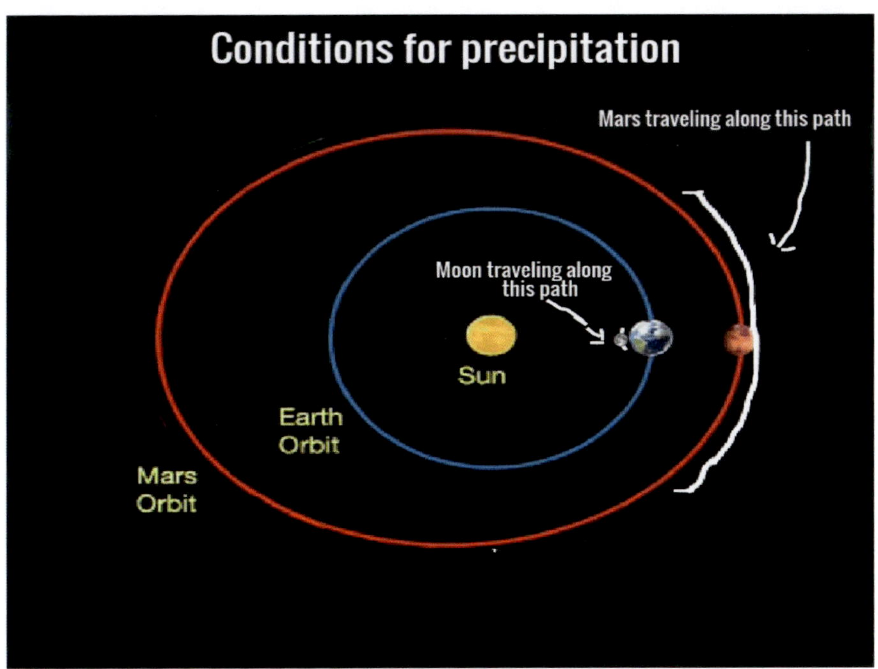

With these narrowed down, we can delve into the other two variations which can also be applied to precipitation science, which involve a close conjunction between the moon and Mars. If the moon is traveling in front of the earth, while Mars is traveling behind the sun, then both bodies in conjunction would be pulling earth's axial tilt toward the sun, exposing the earth to more sunlight and warmth. Here we can make the assumption that the warmer temperatures that result could lead to precipitation as the warm front mixes with the less warmer air, which could lead to the break down of water vapors. Here is an example of this close conjunction.

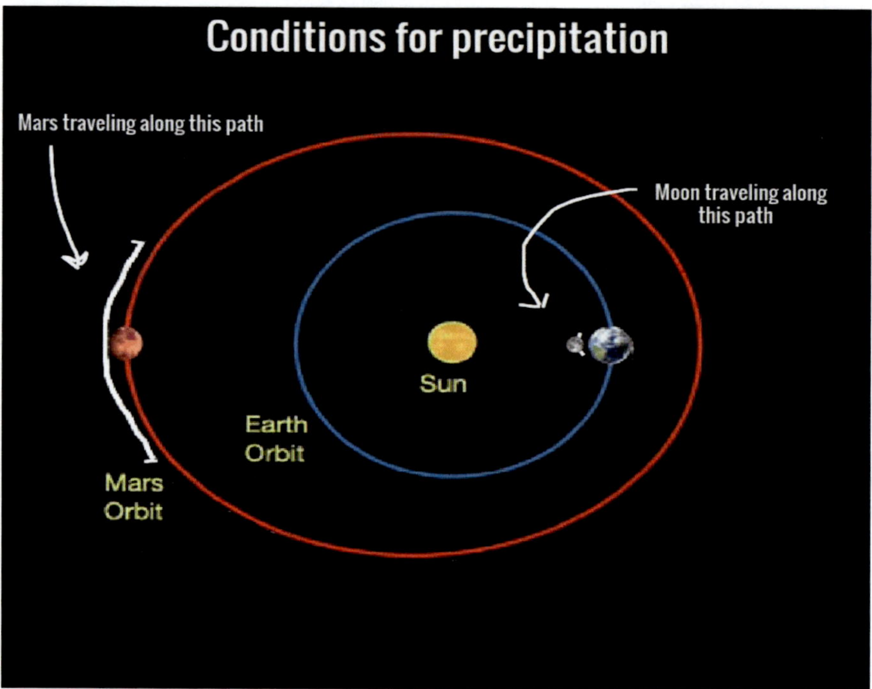

Now onto a visual of the other close conjunction between the moon and Mars, by which the moon is traveling behind the earth, while Mars travels in front of the sun, relative to earth's view point. Both bodies would be applying a gravitational pull upon the earth's tilt, bringing the earth away from the sun and exposing earth to cooler temperatures. Should the resulting cool front temperatures mix with the less cooler air, a break up of water vapors can occur and precipitation can result. Here is a visual of this scenario

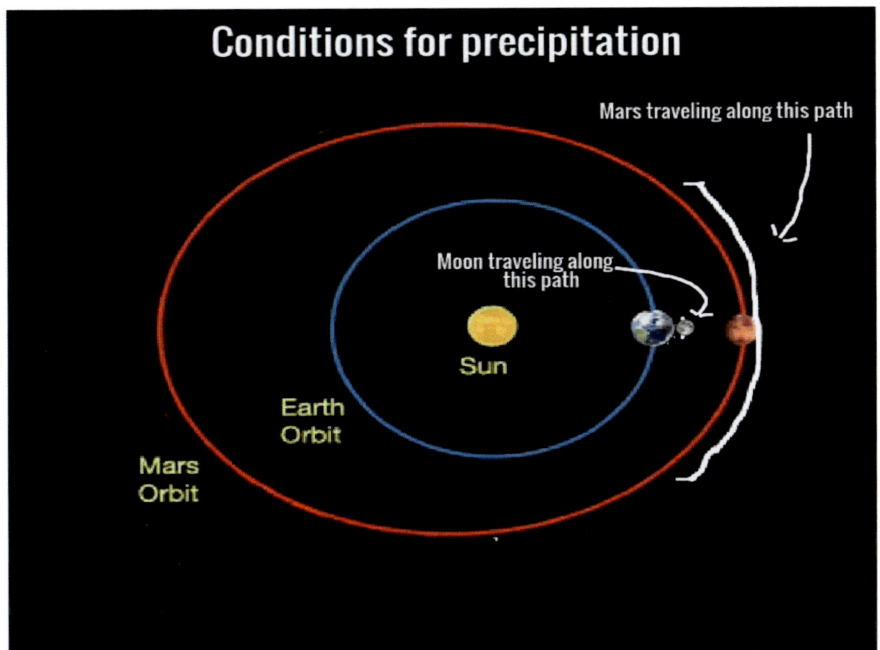

We have now thus far laid out a theoretical framework that could alow us to predict temperature perturbations leading to rainfall by way of both Mars's and the moon's gravity acting upon the earth, in either shifting the earths axial tilt towards and/or away from the sun. However, since this paper has delved into extreme events, as elucidated in the first two sections regarding Gaza rocket fire and Stock market crashes, we should continue on that subject matter and investigate the extreme precipitation events. Much like escalated rocket attacks from Gaza and stock market crashes, we should find a similar theme of Mars being within 30 degrees of the lunar node being a precipitating factor that could trigger extreme precipitation events. Mars within 30 degrees of the lunar node has been explained as a mechanism by which the planet Mars applies a gravitational pull on the orbital path of the moon, stretching it, such that it incrementally brings the moon's orbit further away from earth, a factor that would have a destabilizing effect on earth's wobble, which would expose to earth to wilder temperature fluctuations. If we apply this dynamic to weather events, we can surmise that the scenario could cause major temperature perturbations that can condense water vapors absorbed by the air, triggering rainfall. The moon is factored into this since it is the component that prompts short term temperature perturbations. Keep in mind that we are attempting to explain extreme events. Here is a visual of how the configuration works. This first example is an extreme precipitation event in the Middle East, which occurred in 1979 from October 20th to October 23rd. Fifty people died, and 66000 were affected. Observe the chart and notice that Mars was within 30 degrees of the lunar node and applying the aforementioned gravitational factors. Mars is also behind the sun relative to earth, so it had presumably been a warmer winter.

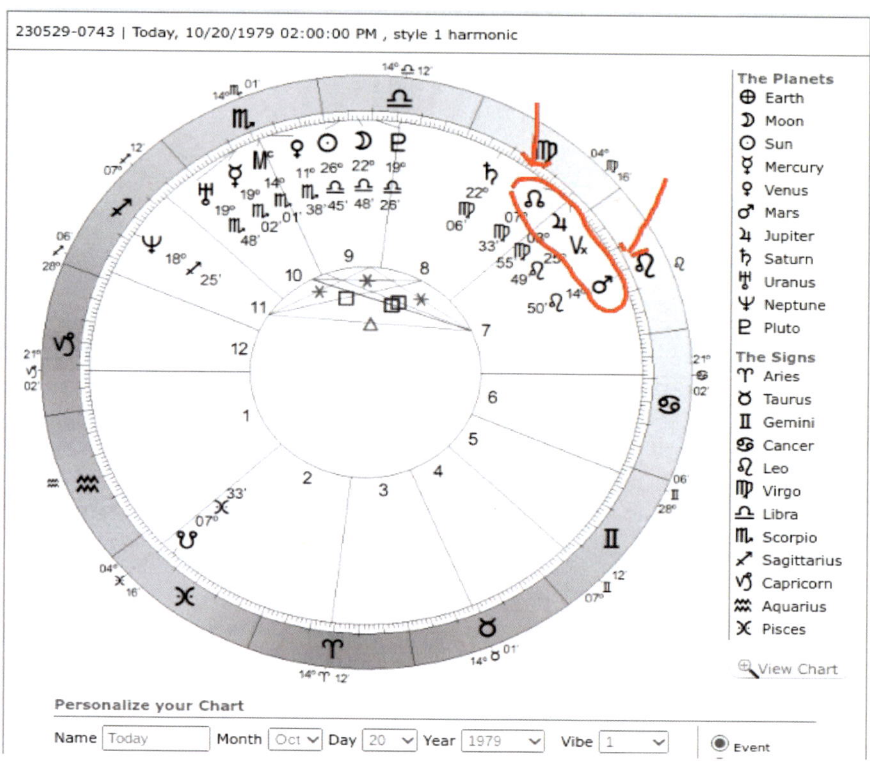

So now the perturbation was triggered by the moon. But pay close attention. I discovered a pattern that indicates that extreme precipitation events can be triggered by right angles between Mars and the moon, if either mass within 30 degrees of the lunar node. So if Mars is within 30 degrees of the lunar node, the temperature perturbation and corresponding precipitation will be triggered when the moon forms a near right angle to the position of Mars. Likewise, if the moon is within 30 degrees of the lunar node, the temperature perturbation can be triggered if the moon is already forming near a right angle to Mars. The former is happening here-Mars is within 30 degrees of the lunar node, while moon at a near right angle to Mars is triggering the temperature perturbation required for extreme precipitation. Here is the visual

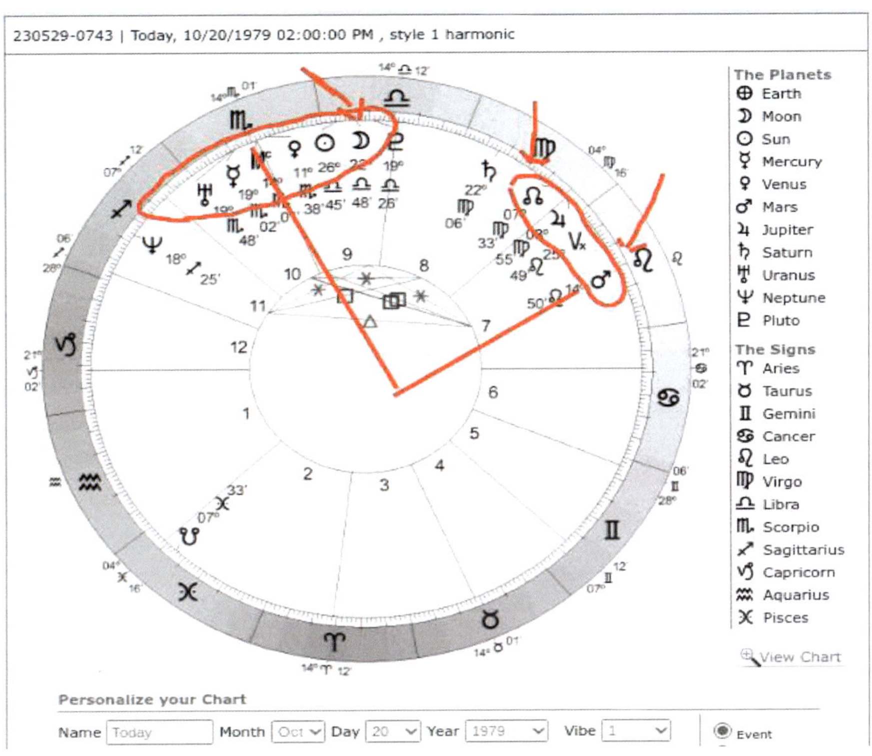

Here is how this configuration appeared in the sky

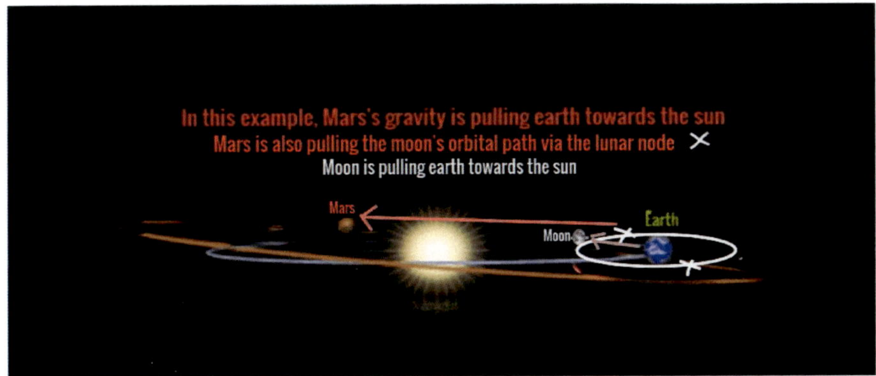

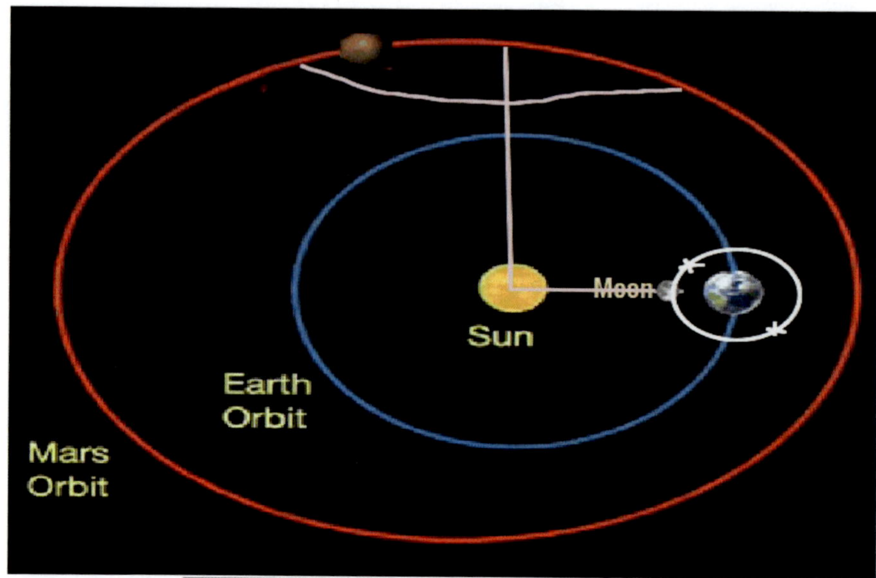

On May 13, 1982, a massive storm caused flooding in the Middle East. Here is the chart, notice that we have a similar dynamic as the first graphic, but this time where the moon is within 30 degrees of the lunar node forming a near right angle to Mars.

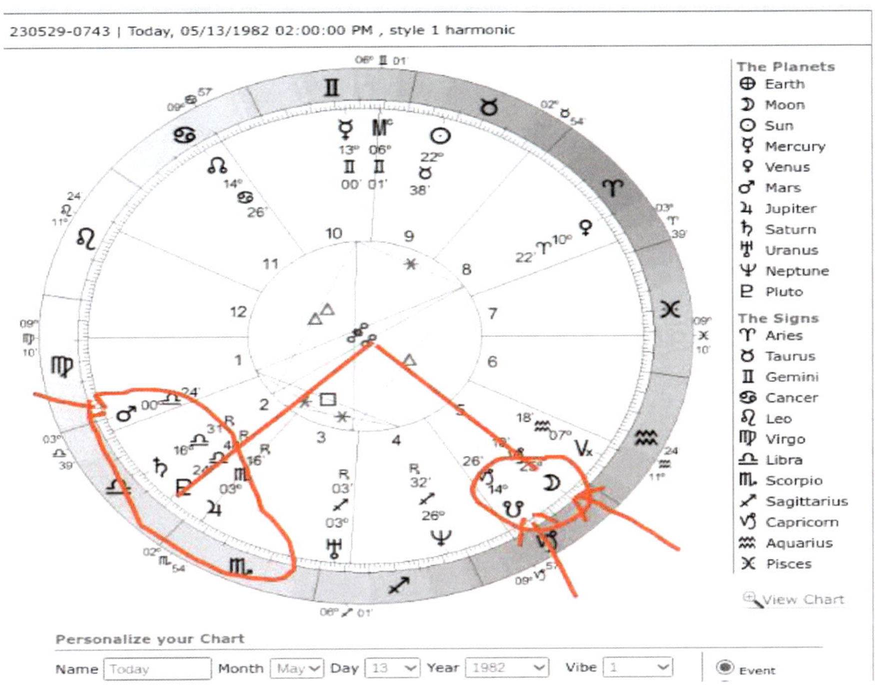

Here is how this configuration appeared in the sky on that day

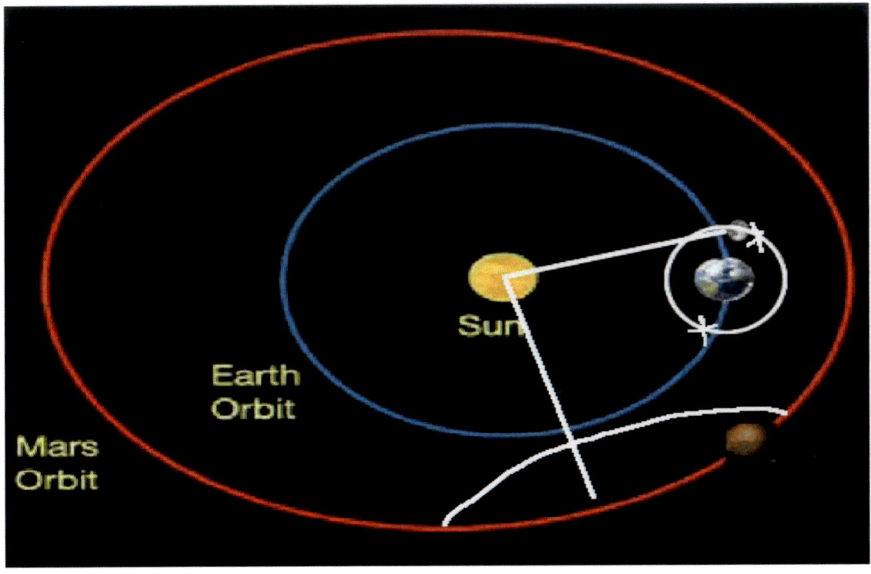

Notice that Mars was within the bounds of the point marking off the right angle created between the configuration of Mars and the moon.

Here is the chart for the storm that occurred on October 16 1987 and affected Egypt and Jordan with flooding, which led to 39 casualties.

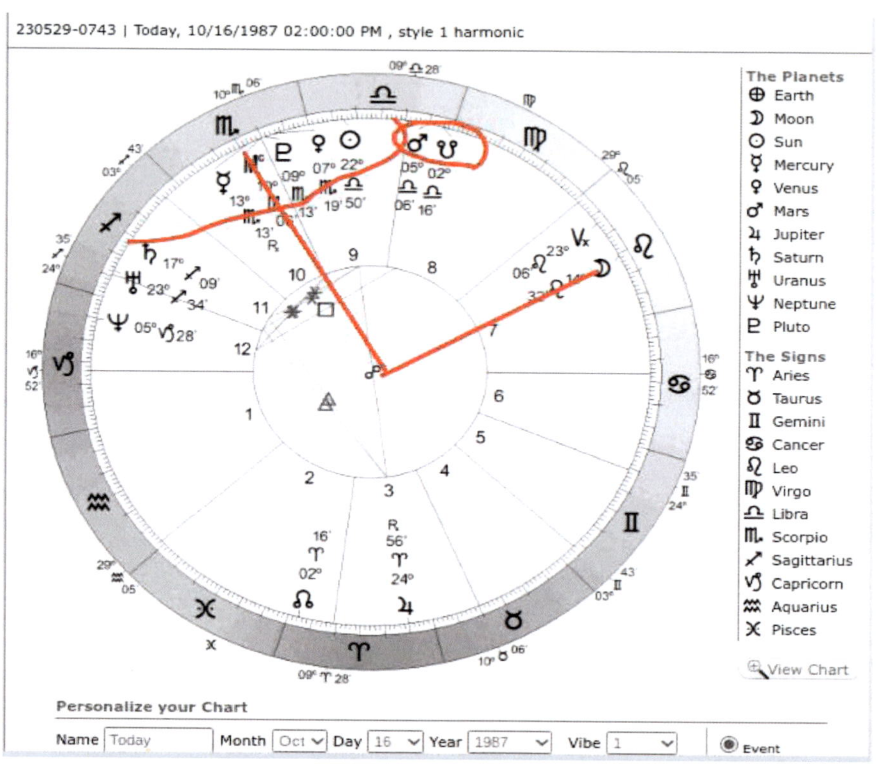

Mars is within 30 degrees of the lunar node and nearly forming a right angle with the moon, but slightly off at the time the chart was calculated. The moon would have been within the assigned range hours earlier. Here is how the configuration appeared in the sky that day

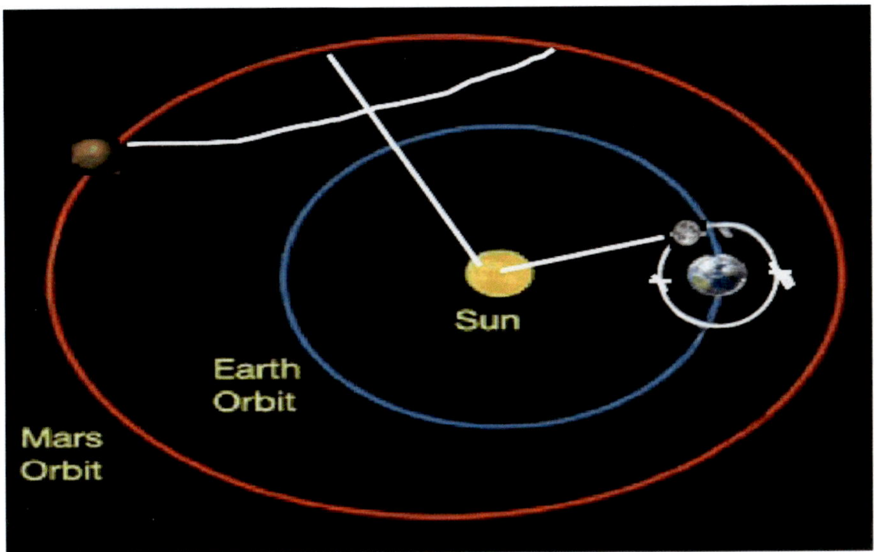

Another heavy precipitation date for Egypt, causing floods occurred on October 16 1988. Here is the astrology chart showing the position of Mars, moon, and the lunar node. Once again Mars was within 30 degrees of the lunar node and making a right angle with the moon

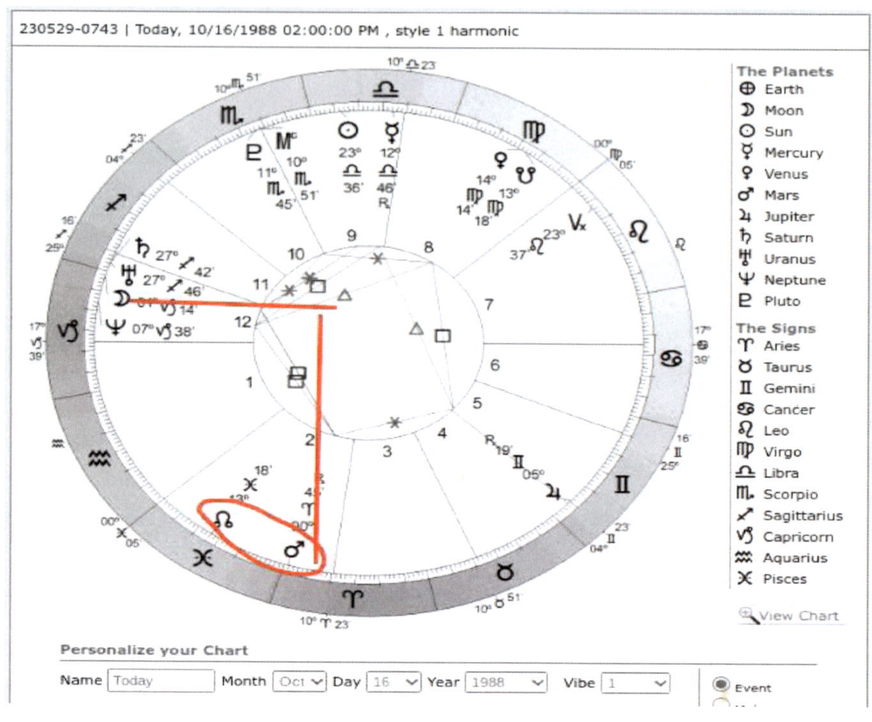

Here is how the configuration appeared in the sky that day

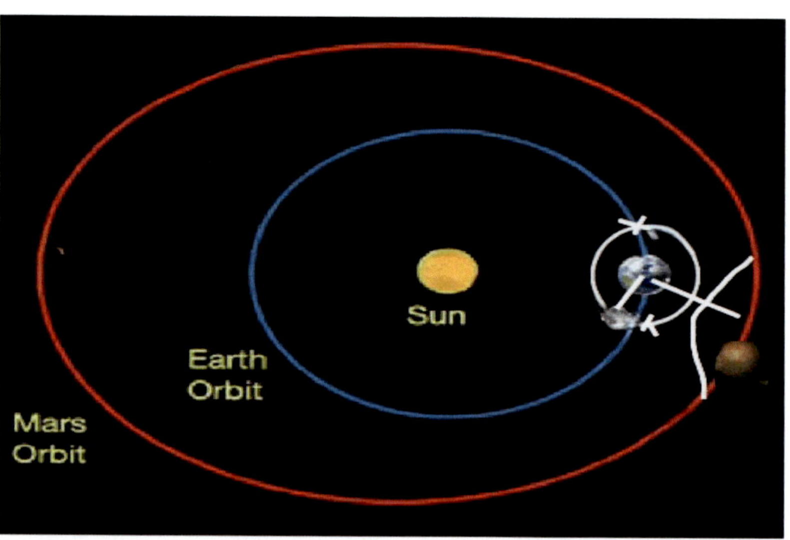

In the sky, the configuration forms a right angle

Another major precipitation event in the levant occurred on October 12, 1991. Here the moon was within 30 degrees of the lunar node and making a near right angle with Mars

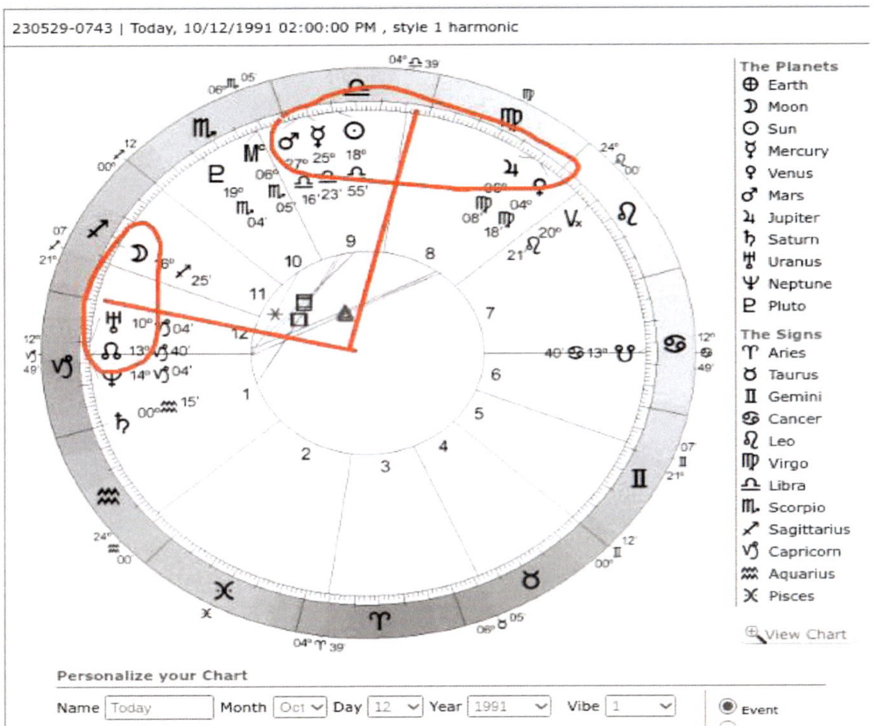

Here is how the configuration appeared in the sky

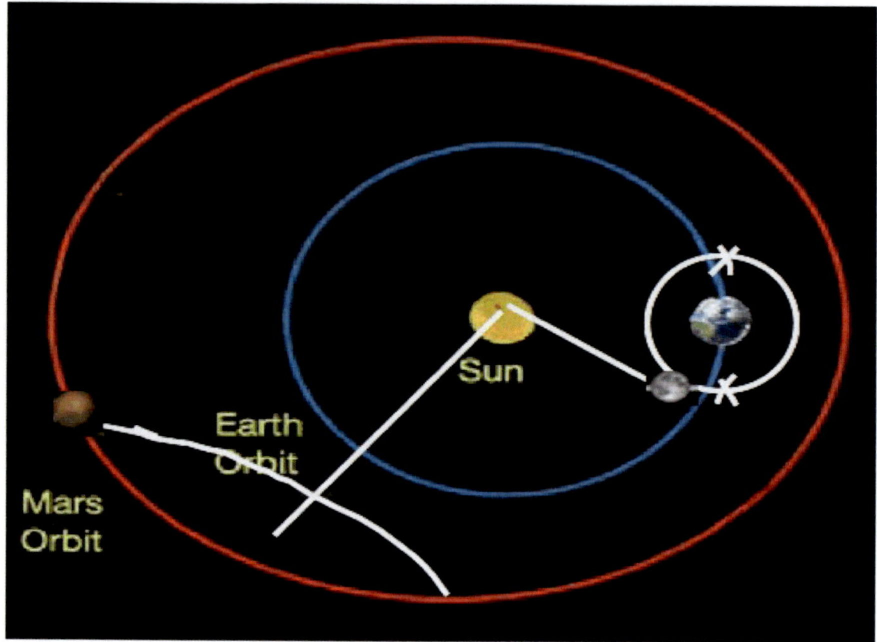

The next major precipitation event in the Levant occurred on December 20 1993. During this time, there were 2 casualties and 10 million in damages in Israel's

In the astrology, Mars was within 30 degrees of the lunar node and making a right angle with the moon, which seems to be a typical configuration for extreme events.

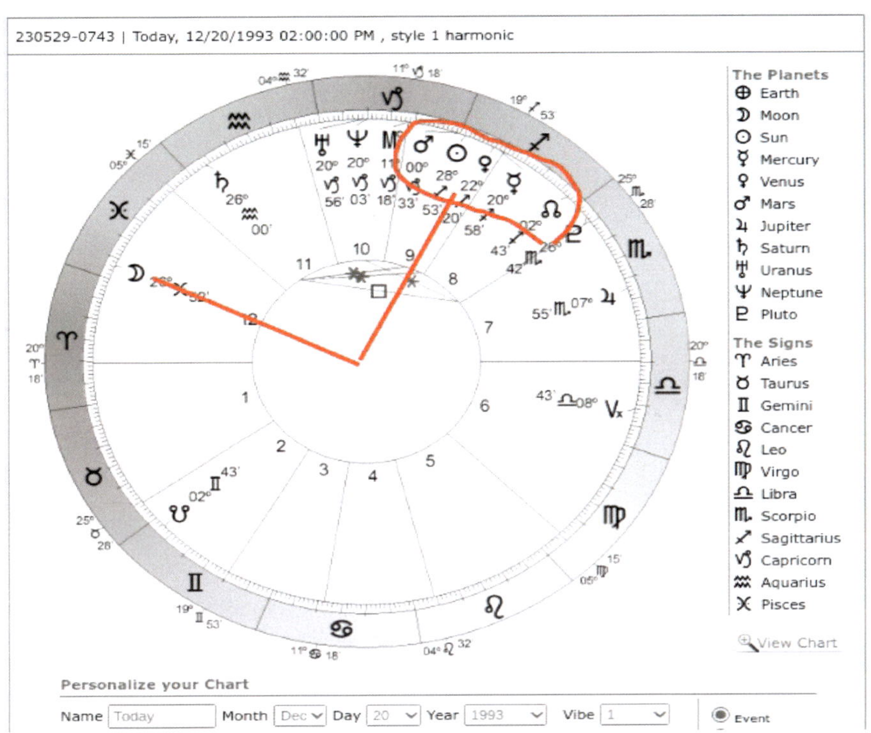

Here is how this configuration appeared in the sky that day

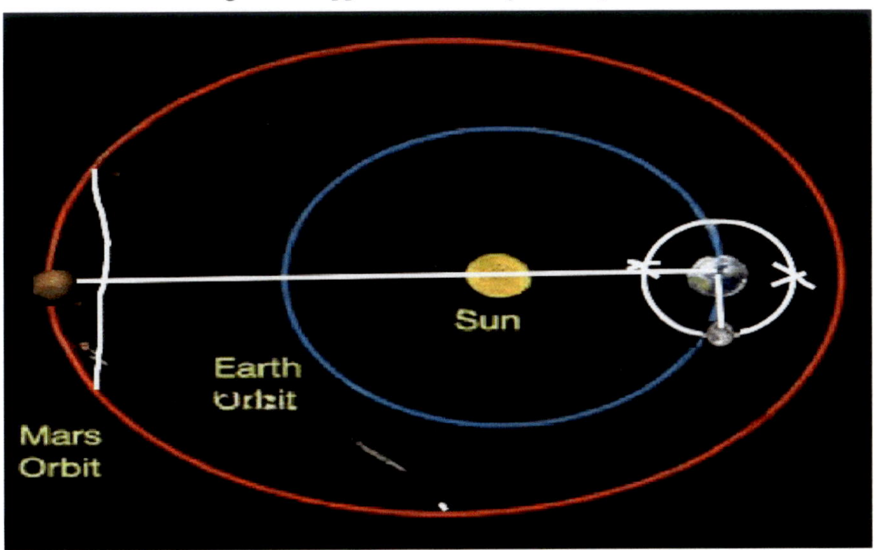

On November 2 of 1994, Egypt was afflicted by extreme flooding which led to 600 casualties and affect 160,000 people, costing 140 million in damages.

During this time, the moon was within 30 degrees of the lunar node and making a right angle with Mars. So once again we see this common pattern in extreme events, with either Mars or moon being within 30 degrees of the lunar node and forming a right angle to the other.

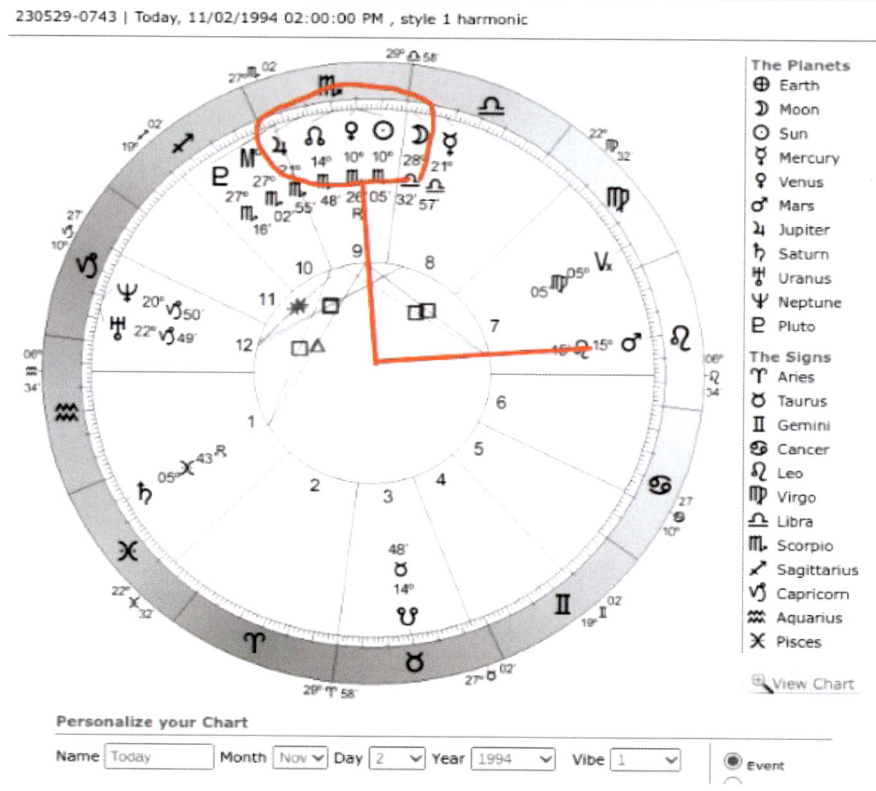

Here is how that configuration appeared in the sky that day

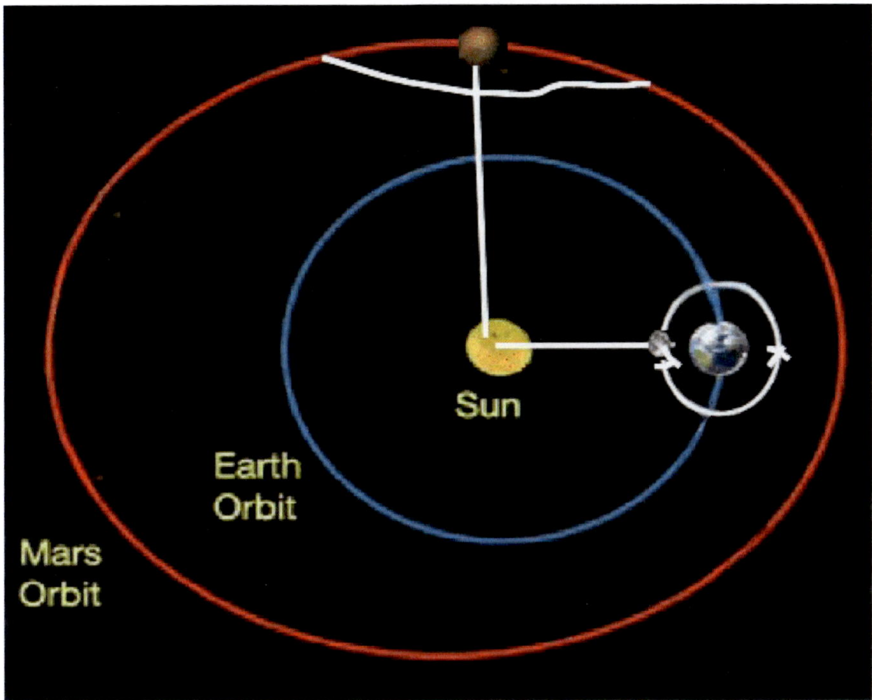

From November 13th to November 18th of 1996, a torrential rains in Egypt resulted in 12 casualties, with 260 people affected by floods. Mars had just began its phase of going within 30 degrees of the lunar node and made a right angle with the moon. Here is the astrology chart

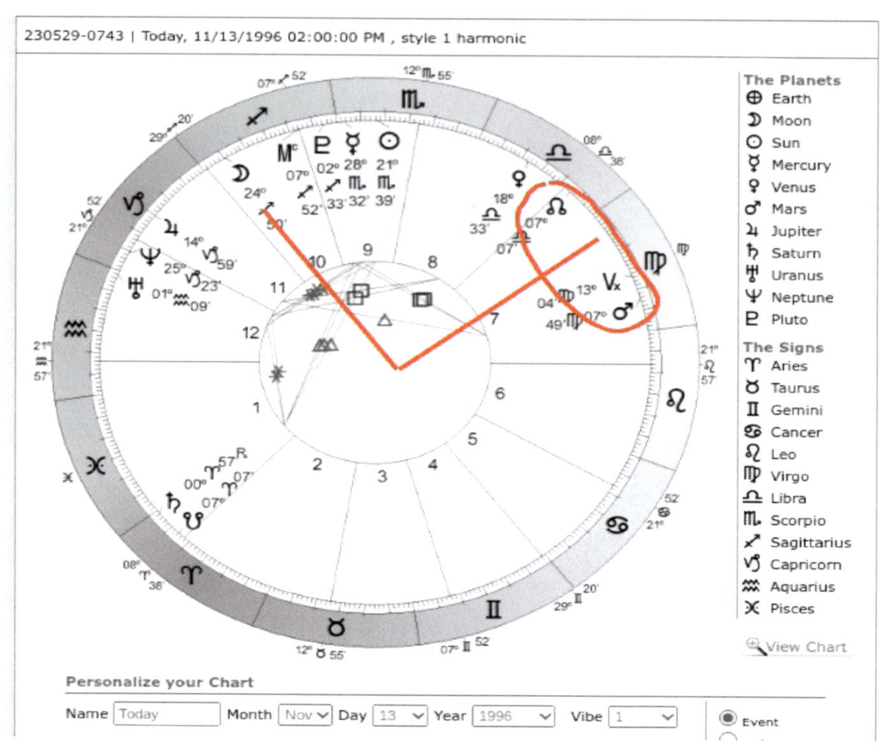

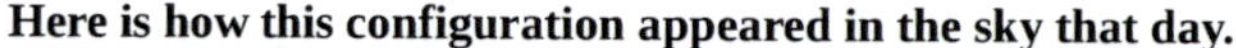

Here is how this configuration appeared in the sky that day.

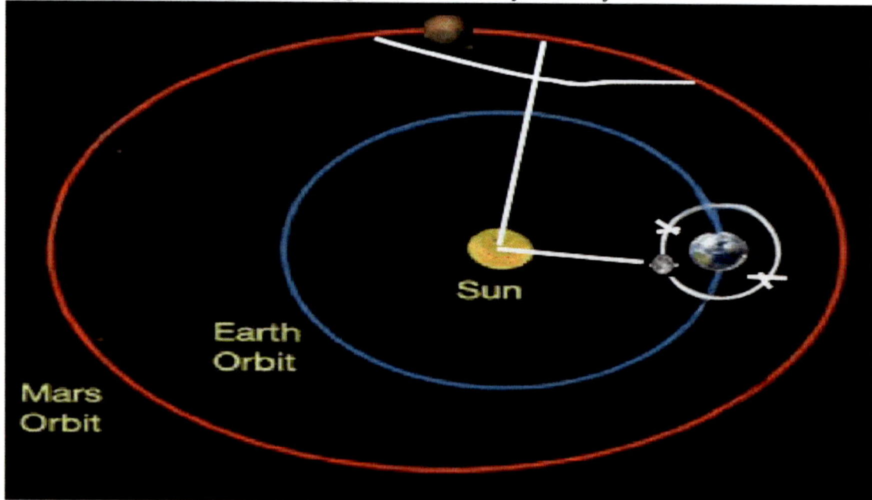

On October 17 1997, heavy rains afflicted Egypt, Israel and Jordan. There were 15 casualties in Israel, Egypt and Jordan, with over 40 million dollars worth of damage. Here is the astrology chart. Here is an example where neither Mars nor the mon were within 30 degrees of the lunar node. This is an example where the moon and Mars where in opposition with each body tugging on Earth's axial tilt, which likely created a temperature perturbation. This is an example of a dynamic by which routine rainfall could be predicted

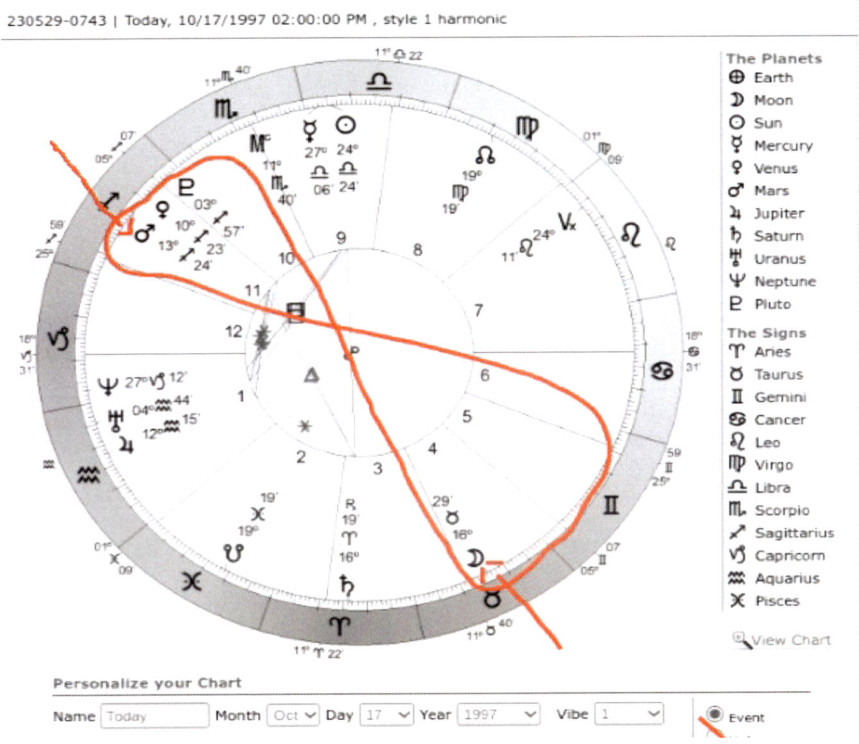

Here is how this configuration appeared in the sky

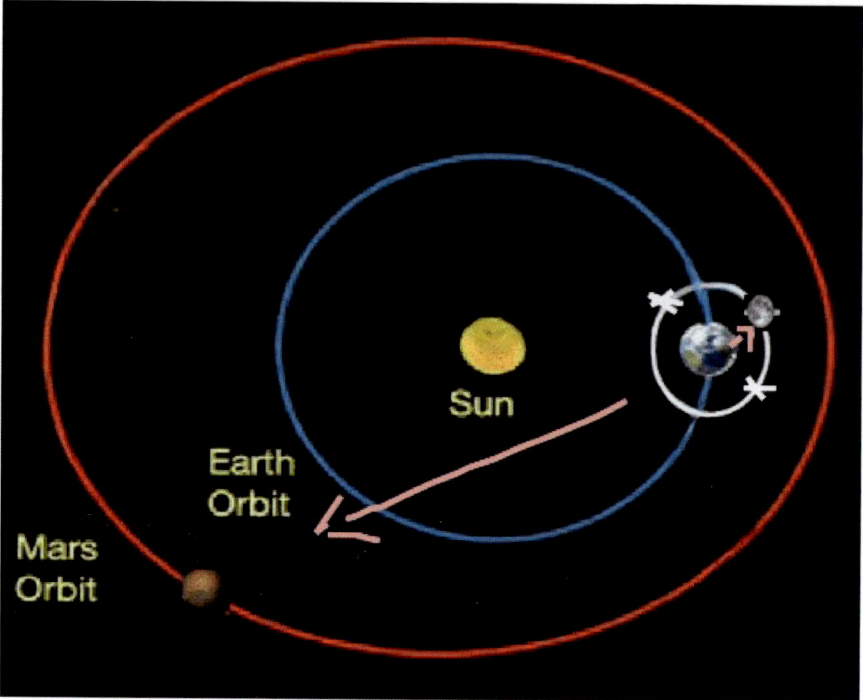

The best explanation as to why Mars within 30 degrees of the lunar node, forming a right angle with the moon, is a catalyst for extreme precipitation events could be due to how this configuration indicates that the moon is traveling along its orbital path furthest from the ecliptic plane. This is not to be confused with apogee and perigee when the moon is respectively furthest and closest to earth in its orbital path. The moon's orbit around the earth is tilted five degrees from the ecliptic and only meets with the ecliptic at the lunar nodes. Yet, during perigee(moon closest to earth) and apogee(moon furthest from earth), the moon is very close to the lunar nodes. So in this regard, we have to observe the moon relative to the ecliptic plane and why its proximity from such is a factor contributing to temperature perturbations and rainfall. We can surmise that when the moon is furthest off the ecliptic plane as Mars gets closer to the lunar nodes, temperature perturbations occur as a consequence of the moon's waning gravitational pull on the earth during this period, allowing Mars to exert its gravitational influence with less opposition from the moon. This could bring in humidity and moisture that would immediately precipitate if it merges with cooler air, assuming this happens during the winter.

This next chart is for January 22 2005. Between the 22 and 27, torrential rain led to 29 casualties in the Middle East. Here is the chart. Mars and the moon are in opposition

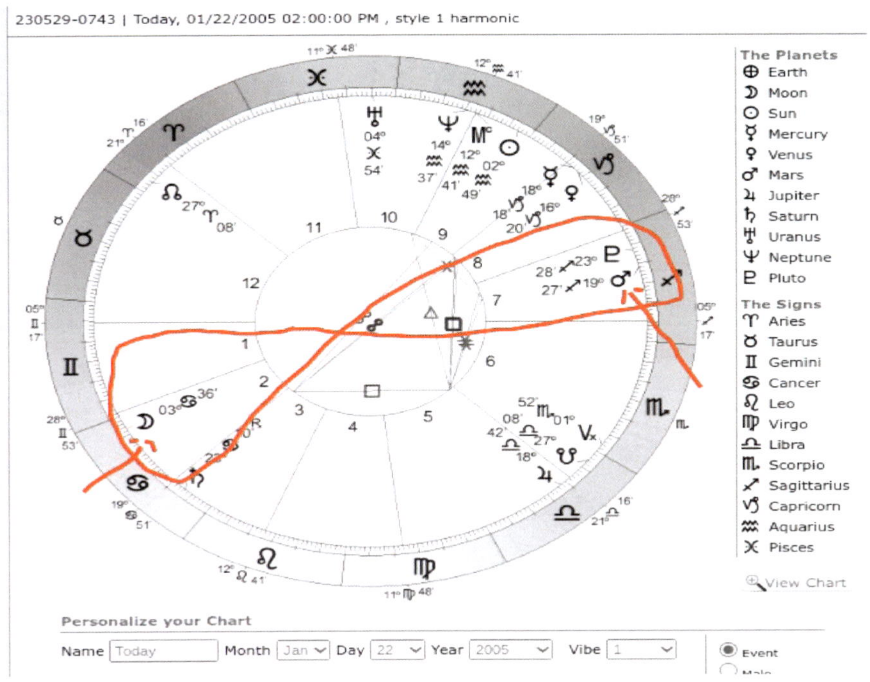

Here is how this configuration appeared in the sky

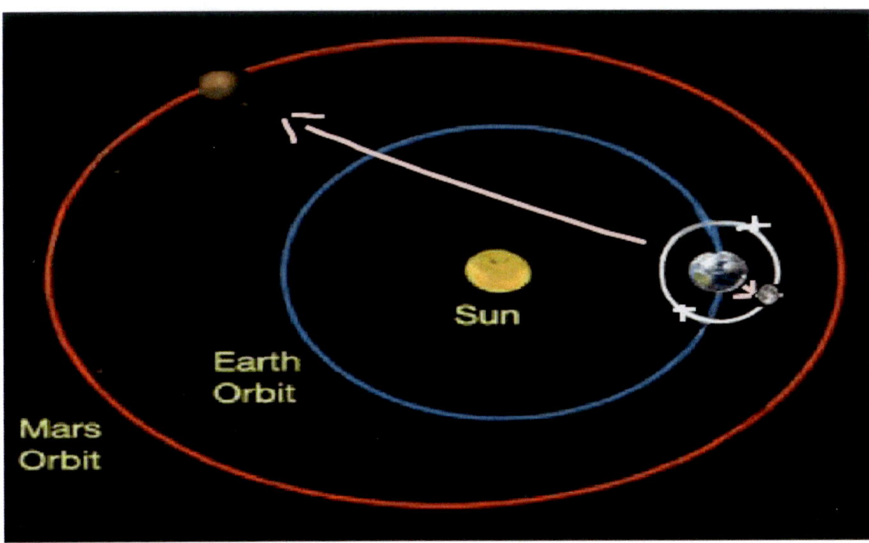

This next chart is for November 25, 2009, a day which led to massive flooding in Saudi Arabia, leading to 122 fatalities. 10,000 people affected, with an estimated 900 million dollars in damage. Mars is within 30 degrees of the lunar node, but the moon is not forming the angle expected for such an event like this one. The moon is opposite Mars, exerting an opposing effect on Mars's gravitational pull.

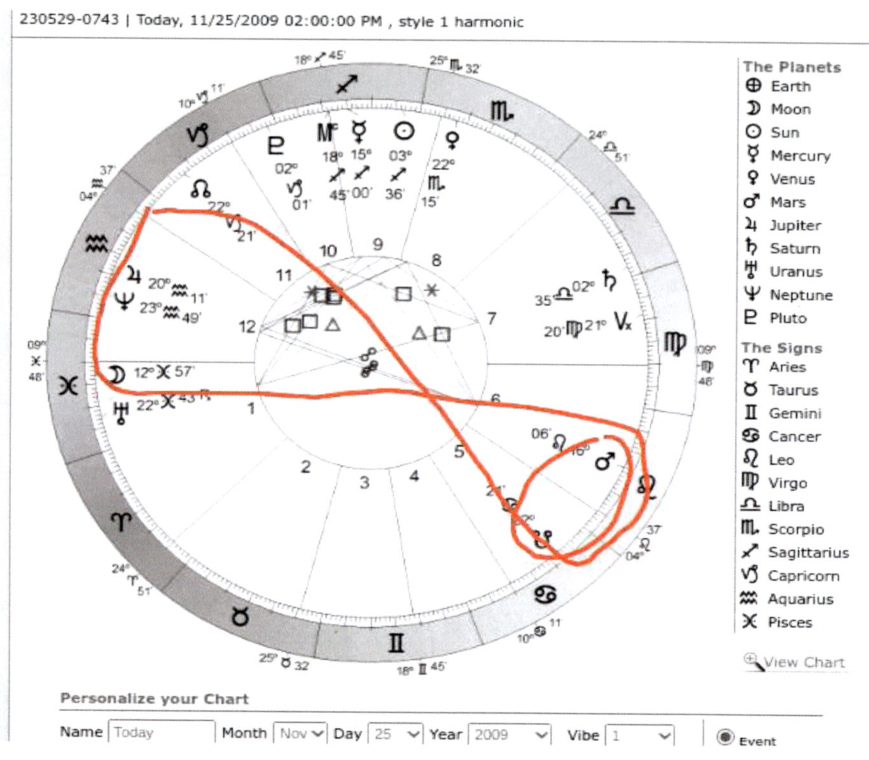

Here is how this configuration appeared in the sky on this day

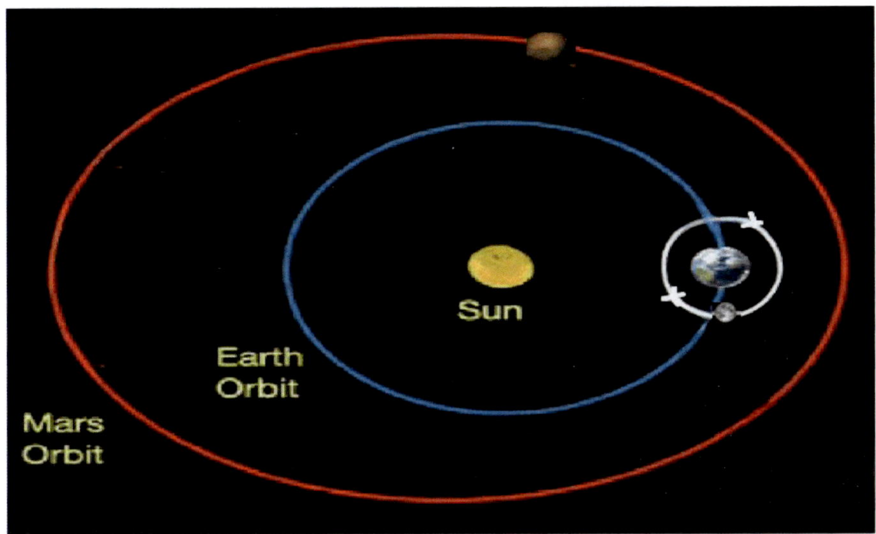

This is the chart for May 2, 2013, when rainfall and flooding led to 20 casualties in the Middle east. This chart shows Mars within 30 degrees of the lunar node forming a right angle with the moon.

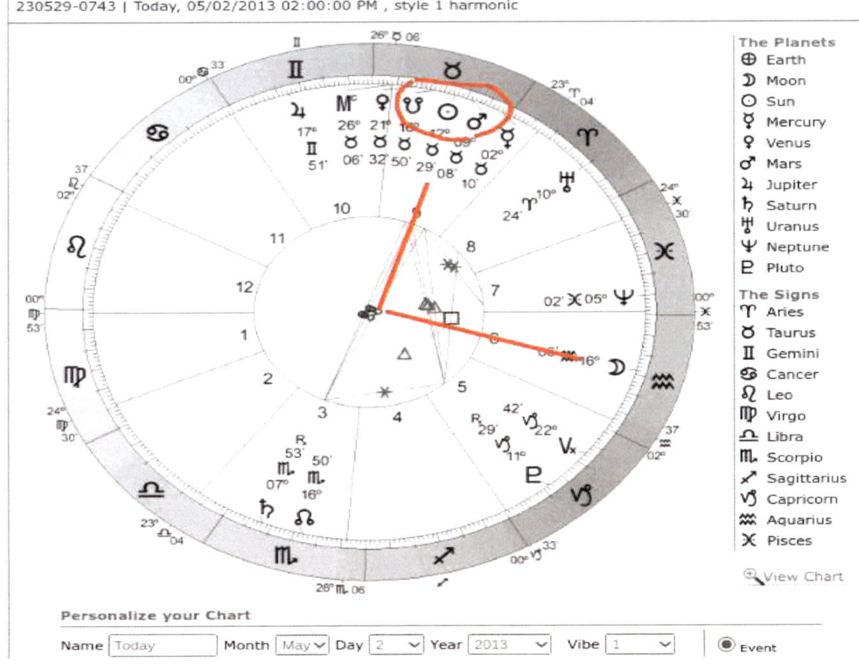

Here is how this configuration appeared in the sky

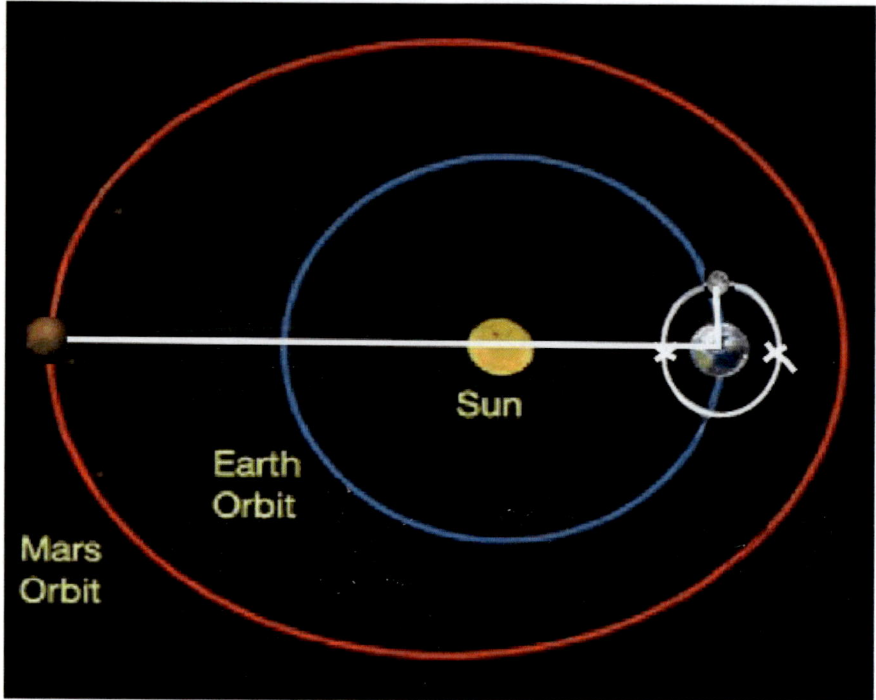

In 7 of 12 of the listed charts for heavy precipitation in the Middle East, Mars was within 30 degrees of the lunar node. Based on that alignment, farmers in the Middle East could use it to develop decisive protocols on how to efficiently allocate water resources and commence fertilizer and cropping activities.

Here are examples of four major storms and floods in last five years in the Middle east.

Here is the chart for March 12, 2020, during which major rainfall and floding struck the Middle East. Nine countries were affected—Egypt, Jordan, Israel, Syria, Lebanon, Turkey, Saudi Arabia, Sudan, Iran and Iraq. Mars was within 30 degrees at this time, making a right angle to the moon. This was Egypt's worst storm since 1979, during which Mars was also within 30 degrees of the lunar node.

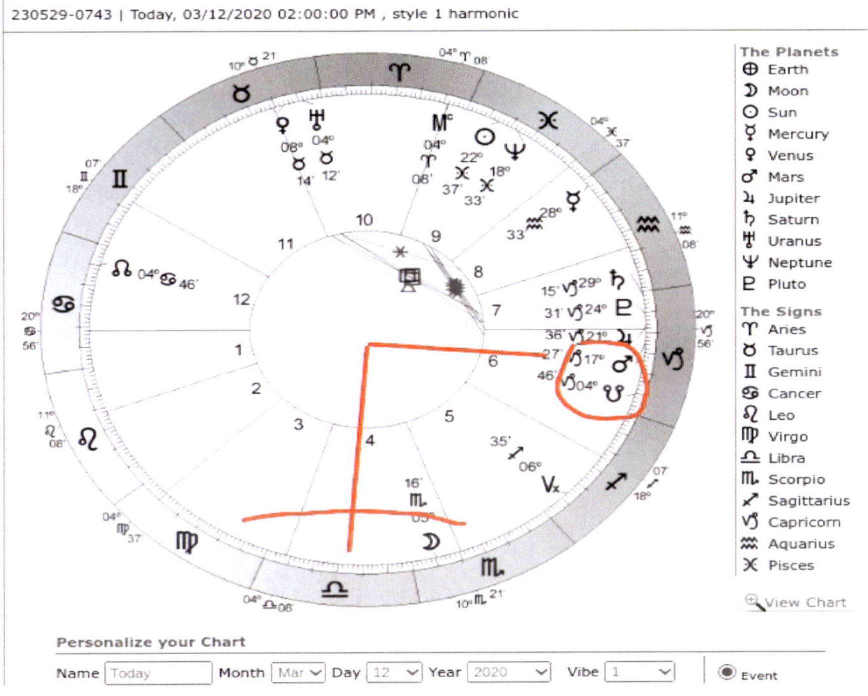

Here is the chart for July 26 2022, during which the United Arab Emirates experienced record breaking rainfall.

Once again, on this day Mars was within 30 degrees of the lunar node, forming a near right angle with the moon at the start. Within hours, the moon would be within the right angle zone

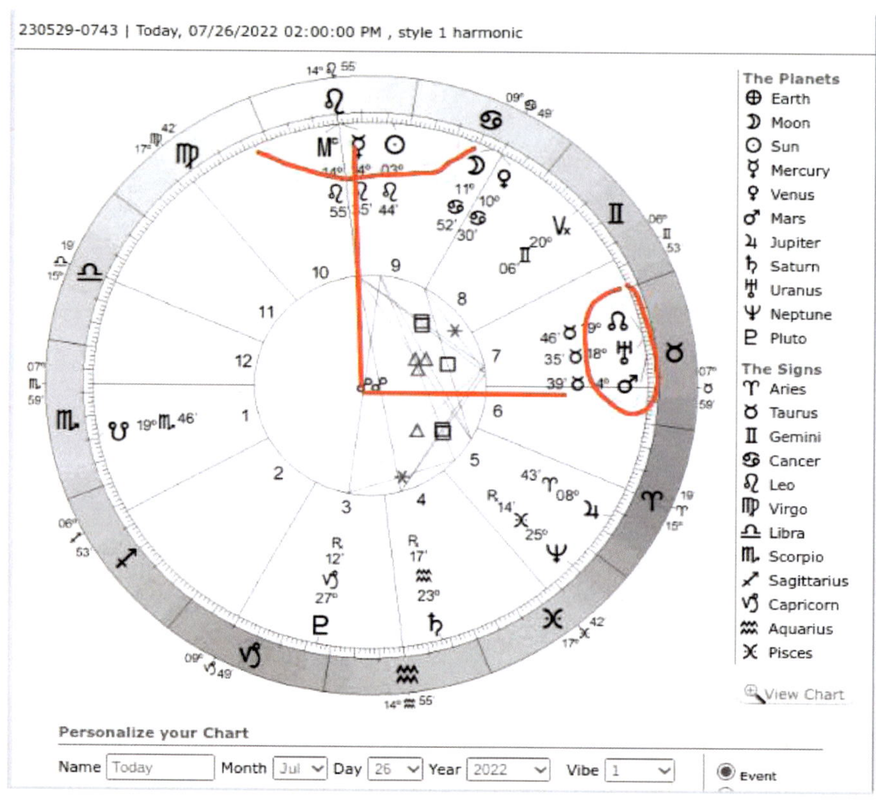

Here is the chart for the 2023 Libya floods, brought on by Storm Daniel, which struck Libya on September 10th 2023. On this day, Mars was within 30 degrees of the lunar node, forming a right angle with the moon

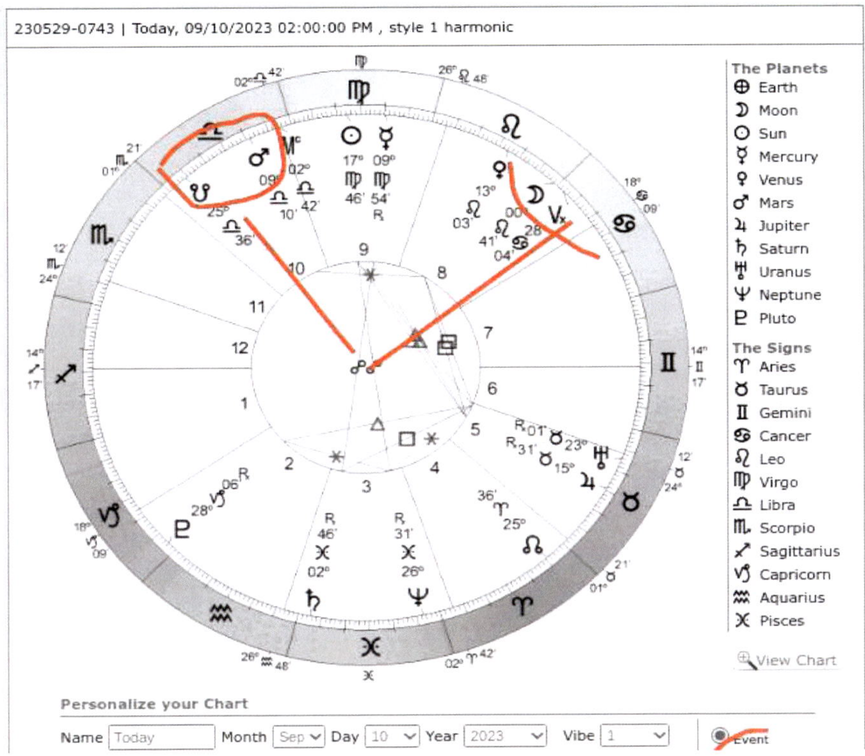

Here is the chart for the United Arab Emirates floods in April of 2024. Heavy rains struck the UAE on April 14th 2024 and caused major flooding. The United Arab Emirates, Oman, Iran, Bahrain, Qatar, Saudi Arabia, Yemen were all affected. This was a record breaking event for the UAE. Once again, Mars was within 30 degrees of the lunar node, forming a right angle with the moon when the storm made landfall there. This was a record breaking event for the UAE

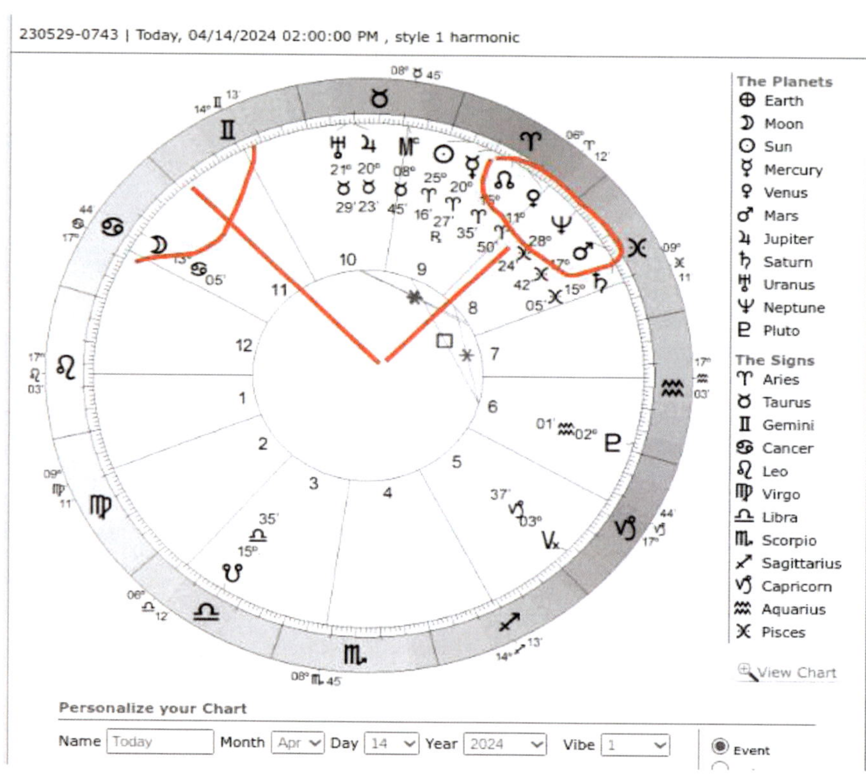

The only extrapolation we can gather for this data is that Mars within 30 degrees of the lunar node may be responsible for above average rainfall in a given season. Here, we can devise a system that could predict heavy rainfall and thus help everyone in the Middle East with emergency response protocols and agricultural timing related to crop growth and development. In irrigated agriculture, the amount of rainfall determines the amount of irrigation water and its consumption time. Rainfall-based systems look to the timing of rainfall to determine crop growth. This also translates into the timing of fertilizer, herbicide and pest control applications. Rainfall is also key to timing harvest operations for post-harvest activities. Forecasting weather events helps in planning farm tasks, planting or not, determining whether to irrigate or not to use fertilizer, transportation and storage of food grains, and measures to protect livestock. In general, a successful weather forecasting system contributes to the decision-making process of agricultural practices

Keep in mind that the premise for the Mars factor was affirmed in 2024, when scientists began to hypothesize that Mars influences Earth's climate and ocean tides.

Here is an article from Science.org

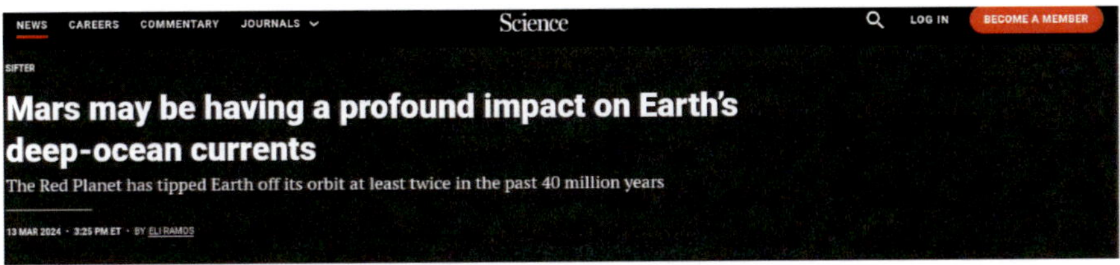

"The moon causes tides, but it's not the only celestial body that affects Earth's water. Mars' gravity affects our planet's deep ocean currents, according to a study published this week in the journal Nature Communications."

Other papers support the hypothesis that Mars must have some influence on Earth. In this section, I combined this dynamic with the scientific premise that the moon has influence over the amount of rainfall via its gravitational pull on earth's atmosphere.

On the next page is an example(sources used) of the dates when the Middle East was exposed to heavy rains, floods and human casualties. The dates are taken from a study that examined the dynamics of extreme rainfall events in the Levant and the Middle East. Source: Extreme Precipitation Events in the Middle East: Dynamics of the Active Red Sea Basin AJ de Vries, E. Tyrlis, D. Edry, S. o . Krishak, B. Steele, J. Lilyfeld. First published: 12 June 2013 https://doi.org/10.1002/jgrd.50569

Nr.	Years and Months	Days	Sources of Motivation [a]	Societal Impact	Case Studies
1	Oct 1979	20–23	1,2	50 casualties, 66,000 people affected, and US$ 14 M damage in Egypt (flood) [b]	
2	May 1982	13			
3	Oct 1987	16–18	1,2	30 casualties in Egypt (storm on 17 Oct) and nine casualties in Jordan (flood on 16 Oct) [b]	
4	Oct 1988	16–19	1		
5	Oct 1991	12–14	1,2,3		Greenbaum et al. [1998]
6	Dec 1993	20–23	3	two casualties and estimated damage US$ 10 M in Israel [c]	Ziv et al. [2005]
7	Oct 1994	10	1,2		

Nr.	Years and Months	Days	Sources of Motivation [a]	Societal Impact	Case Studies
8	Nov 1994	2–4	1,2,3	600 casualties, 160,660 people affected, and US$ 140 M damage in Egypt (flood, 2–8 Nov) [b]	Krichak and Alpert [1998], Krichak et al. [2000]
9	Nov 1996	16–18		12 casualties and 260 people affected in Egypt (flood, 13–18 Nov) [b]	
10	Oct 1997	17–19	1,2,3	15 casualties and US$ 40 M damage in Israel (flood from 17 to 19 October), four casualties, and US$ 1 M damage in Egypt (flood, 18–20 Oct) and two casualties and US$ 1 M damage in Jordan (flood, 18–20 Oct) [b]; at least six casualties in Egypt, nine in Israel, and two in Jordan [c]	Dayan et al. [2001]
11	Nov 2003	23–25			
12	Oct 2004	28–29	3		Greenbaum et al. [2010]

Sources

The Dow's Biggest One-Day Drops
Here's where yesterday's drop of 586 points ranks among the
worst drops in the Dow's history:

Date	Close	Change	Percent
9/29/2008	10,365.45	-777.68	-6.98%
10/15/2008	8,577.91	-733.08	-7.87%
9/17/2001	8,920.70	-684.81	-7.13%
12/1/2008	8,149.09	-679.95	-7.70%
10/9/2008	8,579.19	-678.92	-7.33%
8/8/2011	10,809.85	-634.76	-5.55%
4/14/2000	10,305.78	-617.78	-5.66%
8/24/2015	15,873.22	-586.53	-3.56%
10/27/1997	7,161.14	-554.26	-7.18%
8/21/2015	16,459.75	-530.94	-3.12%

Largest daily percentage losses[5]

Rank	Date	Close	Change	
			Net	%
1	1987-10-19	1,738.74	−508.00	−22.61
2	2020-03-16	20,188.52	−2,997.10	−12.93
3	1929-10-28	260.64	−38.33	−12.82
4	1929-10-29	230.07	−30.57	−11.73
5	2020-03-12	21,200.62	−2,352.60	−9.99
6	1929-11-06	232.13	−25.55	−9.92
7	1899-12-18	58.27	−5.57	−8.72
8	1932-08-12	63.11	−5.79	−8.40
9	1907-03-14	76.23	−6.89	−8.29
10	1987-10-26	1,793.93	−156.83	−8.04
11	2008-10-15	8,577.91	−733.08	−7.87
12	1933-07-21	88.71	−7.55	−7.84
13	2020-03-09	23,851.02	−2,013.76	−7.79
14	1937-10-18	125.73	−10.57	−7.75
15	2008-12-01	8,149.09	−679.95	−7.70
16	2008-10-09	8,579.19	−678.91	−7.33
17	1917-02-01	88.52	−6.91	−7.24
18	1997-10-27	7,161.14	−554.26	−7.18
19	1932-10-05	66.07	−5.09	−7.15
20	2001-09-17	8,920.70	−684.81	−7.13

2005
Source: https://www.terrorism-info.org.il/en/18892/

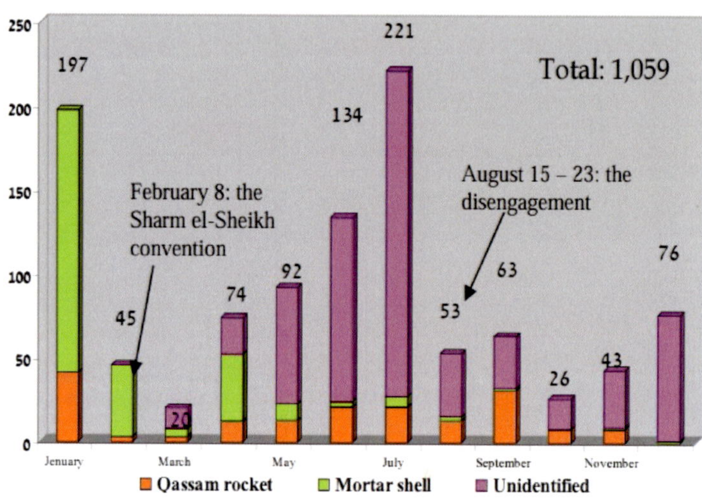

2006
Source: https://www.terrorism-info.org.il/en/18614/

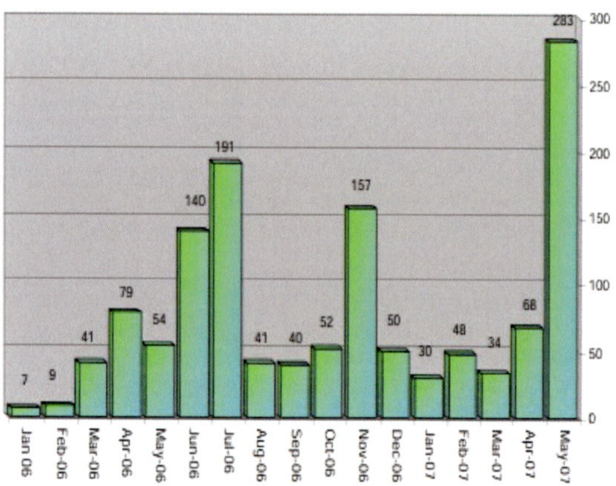

2007
Source: https://www.terrorism-info.org.il/en/18534/

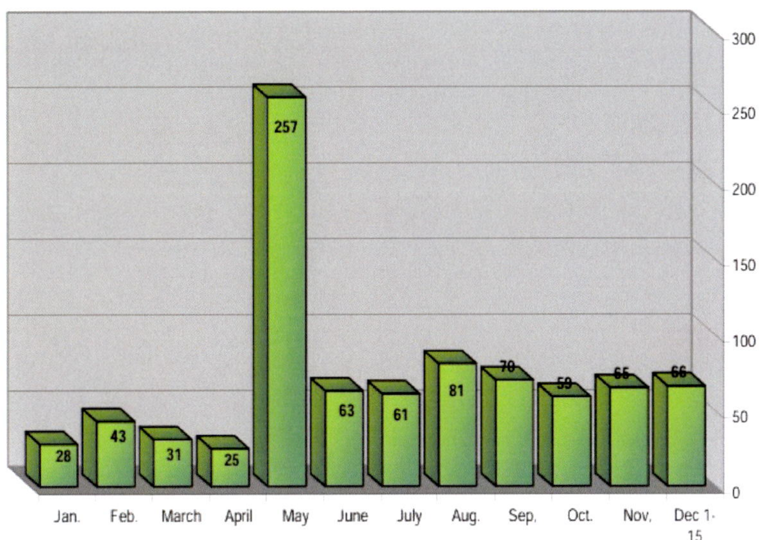

2008
Source:https://en.wikipedia.org/wiki/File:Rock_mort_gaza_2008.JPG

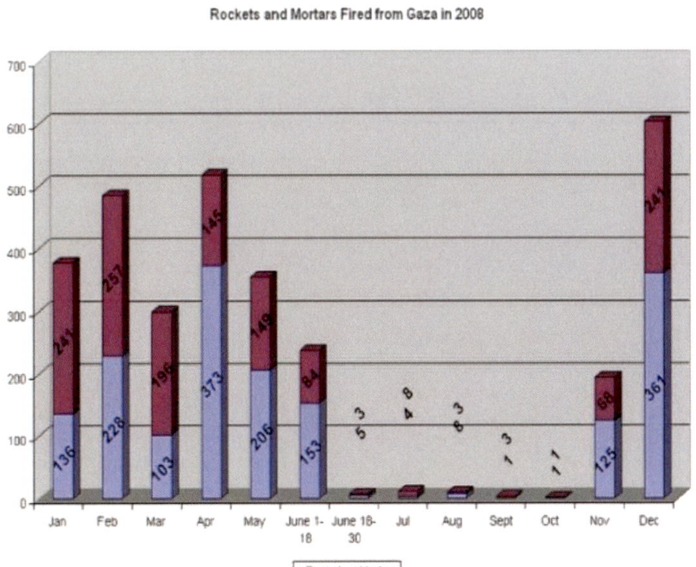

2009

Source: https://www.shabak.gov.il/reports/

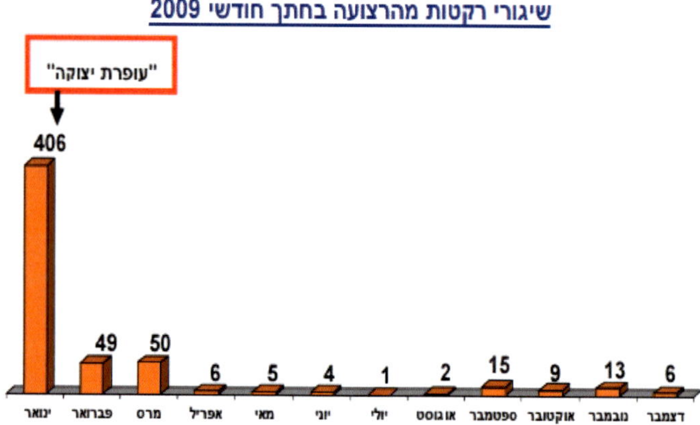

2010

Source: https://www.shabak.gov.il/reports/

שיגורי רקטות מהרצועה בחתך חודשי 2010

סה"כ: 152 שיגורים

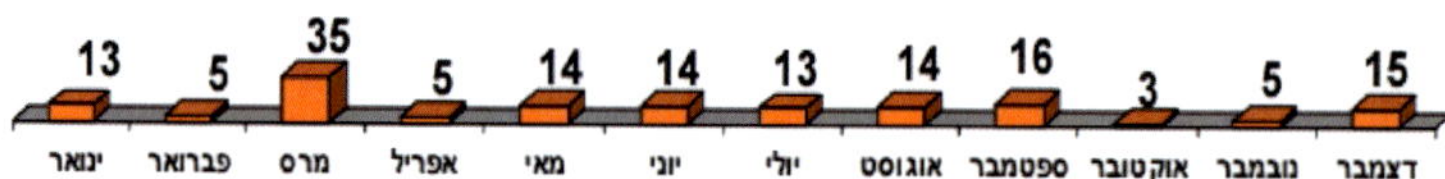

2011
Source: https://en.wikipedia.org/wiki/List_of_Palestinian_rocket_attacks_on_Israel_in_2011

Month	Missiles launched		Effect of missiles		Retaliation by Israel	
	Rockets	Mortars	Killed	Injured	Killed	Injured
January	17	26		4		
February	6	19			1	17
March	38	87		3	9	8
April	87	57	1	6	8	23
May	1					
June	4	1				
July	20	2				2
August	145	46	1	30	4	2
September	8	2				
October	52	6	1	2	12	
November	11	1		1	2	6
December	30	11			4	4
Total	419	258	3	46	40	62

2012
Source: https://en.wikipedia.org/wiki/List_of_Palestinian_rocket_attacks_on_Israel_in_2012

Month	Missiles launched		Effect of missiles		Retaliation by Israel	
	Rockets	Mortars	Killed	Injured	Killed	Injured
January	9	7				
February	36	1			1	1
March	173	19		14	26	
April	10					
May	3					
June	83	11		1		
July	18	9		1		
August	21	3		1		
September	17	8		7		
October	116	55			8	2
November	1734	83	6	45	6	51
December	1					
Total	2,221	196	6	69	41	54

2013

Source: https://en.wikipedia.org/wiki/List_of_Palestinian_rocket_attacks_on_Israel_in_2013

Month	Missiles launched		Effect of missiles		Retaliation by Israel	
	Rockets	Mortars	Killed	Injured	Killed	Injured
January						
February	1					
March	4					
April	17	5			1	
May	1	4				
June	5					
July	5	2				
August	4					
September	8					
October	3	2				
November		5				
December	4					
Total	52	18	0	0	1	0

2014

Source: https://en.wikipedia.org/wiki/List_of_Palestinian_rocket_attacks_on_Israel_in_2014

Month	Missiles launched		Effect of missiles		Retaliation by Israel	
	Rockets	Mortars	Killed	Injured	Killed	Injured
January	22	4				
February	9					
March	65	1		1	1	
April	19	5				
May	4	3				
June	62	3		6		
July	2,874	15[6]	6	34	1,122	7,800
August	950		2	19	540	1,913
Total	4,005	31	8 94	60	1,663	9,713

2015
Source:

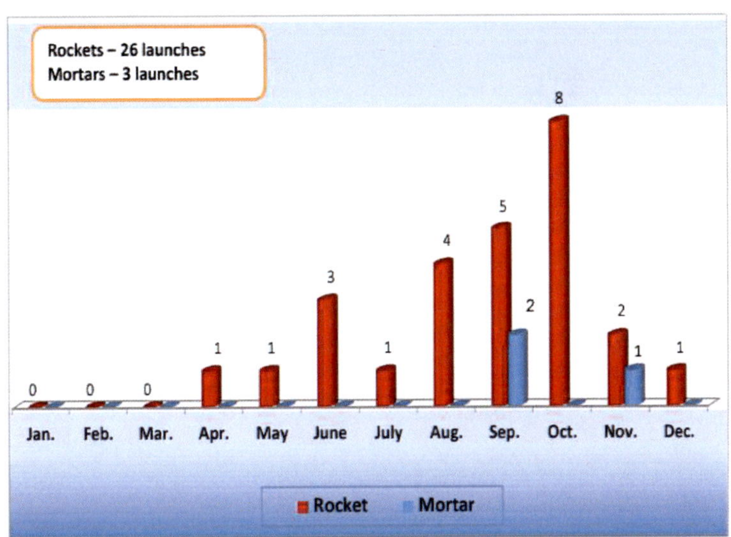

See Jewish virtual library for statistics between 2016 and 2022

https://www.jewishvirtuallibrary.org/palestinian-rocket-and-mortar-attacks-against-israel

In 2023, the data was taken from both

https://www.jewishvirtuallibrary.org/palestinian-rocket-and-mortar-attacks-against-israel

and

Wikipedia
https://en.wikipedia.org/wiki/List_of_Palestinian_rocket_attacks_on_Israel_in_2023

In 2024, the data was taken from
https://www.shabak.gov.il/reports/

and also from news sources about Iran's attack in April of 2024

Rocket data from 2006 Lebanon war
Source: https://en.wikipedia.org/wiki/2006_Lebanon_War

Israeli Military and Civilian casualties in the 2006 Lebanese war[239]

Date	Soldiers			Civilians		Rockets fired on Israel
	Killed	Wounded	Captured	Killed	Wounded	
12 July	8	4	2		2	22
13 July		2		2	67	125
14 July	4	2		2	19	103
15 July		4			16	100
16 July		17		8	77	47
17 July					28	92
18 July		1		1	21	136
19 July	2	15		2	18	116
20 July	5	8			16	34
21 July	1	3			52	97
22 July		7			35	129
23 July				2	45	94
24 July	4	27			17	111
25 July		10		2	60	101
26 July	8	31		1	32	169
27 July		6			38	109
28 July		10			19	111
29 July		7			10	86
30 July		8			81	156
31 July		12			25	6
1 August	3	12				4
2 August	1	41		1	88	230
3 August	4	22		8	76	213
4 August	3	25		3	97	194
5 August	2	70		4	59	170
6 August	12	35		4	150	189
7 August	3	35			12	185
8 August	6	74			10	136
9 August	15	186			36	166
10 August	2	123		2	21	155
11 August	1	76			26	123
12 August	24	131			24	64
13 August	9	203		1	105	217
14 August		37			2	
15 August	2					
Total	119	1244	2	43	1384	3990